Global Millets Production for a Sustainable Future

C. B. Singh • Shilpa Mishra
Ankita Jasmine Lall
Editors

Global Millets Production for a Sustainable Future

Springer

Editors
C. B. Singh
Bundelkhand University
Institute of Economics and Finance
Jhansi, Uttar Pradesh, India

Shilpa Mishra
Institute of Economics and Finance
Bundelkhand University
Jhansi, Uttar Pradesh, India

Ankita Jasmine Lall
Christ Church College
Kanpur, Uttar Pradesh, India

ISBN 978-3-031-88156-5 ISBN 978-3-031-88157-2 (eBook)
https://doi.org/10.1007/978-3-031-88157-2

This Springer imprint is published by the registered company Springer Nature Switzerland AG
The registered company address is: Gewerbestrasse 11, 6330 Cham, Switzerland

We dedicate this manuscript, Global Millets Production for a Sustainable Future, to the world's farmers, scientists, and policymakers, who work tirelessly and with unwavering determination to ensure food security for the global population through sustainable agricultural practices.

To the millions of farmers, whose steadfast dedication transforms soil into a cradle of hope, sowing and nurturing the seeds of a healthier and brighter future.

To the researchers and agricultural scientists, whose ingenuity, commitment, and innovation illuminate the path toward sustainability and resilience in food systems.

To the policymakers, marketers, and NGOs, whose relentless efforts raise awareness and inspire action, championing millets as the cornerstone of nutritional security.

May this work serve as a beacon, fostering collective action for a world where sustainable agriculture thrives, and every individual enjoys the blessings of food security, health, and well-being.

With profound respect and gratitude!

Prof. (Dr.) C. B. Singh
Dr. Ankita Jasmine Lall
Dr. Shilpa Mishra

Foreword

Millets, often referred to as "miracle grains," have captured the attention of agriculturalists, nutritionists, and policymakers worldwide for their extraordinary potential to address some of the most pressing global challenges. As biomedical researcher and an academician with 40 years of experience in promoting sustainable practices, I am deeply honored to introduce this timely and transformative work, *Global Millets Production for a Sustainable Future*.

The renewed focus on millets, particularly in light of the International Year of Millets 2023, marks a pivotal moment in our collective journey toward sustainability. I first encountered the profound ecological and nutritional benefits of millets during my interactions with farmers and researchers working tirelessly to preserve these ancient grains. Their stories and groundbreaking research inspired me to advocate for millets as a cornerstone of sustainable agriculture and food security.

My association with the editors of this volume, distinguished scholars and practitioners, has been a journey of shared passion and commitment. Their expertise, coupled with the contributions of global collaborators, makes this book a unique and invaluable resource. I have had the privilege of working alongside several contributors, including Dr. C.B. Singh and Beatriz Lucia Salvador Bizotto, whose pioneering work on ecological sustainability has laid a strong foundation for reimagining agricultural practices.

This book offers a multidimensional exploration of millets, delving into critical aspects such as production systems, nutritional benefits, and their alignment with Sustainable Development Goals. It highlights innovative case studies, data-driven analyses, and actionable frameworks that will resonate with a diverse audience—farmers, scientists, policymakers, and entrepreneurs alike. The global perspective brought by contributors from Brazil, India, and other regions underscores the universal relevance of millets in tackling climate change, food insecurity, and rural development. From the economic viability of millet farming to its role in improving rural livelihoods, the insights presented here are both profound and practical.

As you turn the pages of this book, you will find a wealth of knowledge and inspiration. Whether you are a farmer seeking sustainable cultivation practices, a policymaker aiming to promote millet-based strategies, or a researcher exploring innovative frameworks, this book will empower you to champion the resurgence of these remarkable grains.

I commend the editors, contributors, and Springer Nature for curating this impactful volume. It is my firm belief that *Global Millets Production for a Sustainable Future* will ignite a movement—one that repositions millets not just as grains of the past but as grains of the future.

Let this book be a beacon of hope and a call to action for sustainable agriculture, nutritional security, and environmental stewardship.

Govind Guru Tribal University
Banswara, Rajasthan, India

K. S. Thakur

Preface

Millets, often celebrated as "miracle grains," have re-emerged as vital crops for achieving global food security, environmental sustainability, and improved nutritional outcomes. The International Year of Millets (2023) provided a platform to highlight the untapped potential of these ancient grains, emphasizing their role in addressing contemporary challenges like climate change, health crises, and rural livelihoods.

This book, *Global Millets Production for a Sustainable Future*, is a collaborative effort that unites contributions from eminent researchers, scientists, and academicians spanning India and the globe. It offers a comprehensive exploration of the multifaceted aspects of millet production, value addition, consumption patterns, and their alignment with Sustainable Development Goals (SDGs). From ecological sustainability in Brazil to millet-based innovations in India, the chapters in this volume reflect rigorous research and on-the-ground experiences, making it a valuable resource for diverse stakeholders.

This book is structured to benefit a wide audience:

- *Farmers* will gain insights into best practices for millet cultivation, its economic viability, and its potential to sustain rural livelihoods.
- *Scientists and researchers* will find valuable case studies, data-driven analyses, and frameworks highlighting millet's resilience, nutritional value, and contribution to sustainable agriculture.
- *Policymakers and NGOs* will uncover actionable recommendations to integrate millet-based strategies into national food security and agricultural policies, fostering both economic and ecological benefits.

I extend my gratitude to the contributors, whose expertise has enriched this work, and to Springer Nature for their commitment to publishing scholarly research that fosters a sustainable future. It is my hope that this book will inspire continued exploration, innovation, and action to secure a healthier and more sustainable world through millets.

Jhansi, Uttar Pradesh, India — C. B. Singh
Kanpur, Uttar Pradesh, India — Ankita Jasmine Lall
Jhansi, Uttar Pradesh, India — Shilpa Mishra
12 January 2025

Acknowledgments

The publication of *Global Millets Production for a Sustainable Future* marks the collective effort of a distinguished group of researchers, practitioners, and policy-makers from across the globe. This book represents a vital contribution to the ongoing discourse surrounding the resurgence of millets as a sustainable and resilient solution to the global challenges of food security, climate change, and agricultural resource management.

We would like to express our sincere gratitude to all the contributors whose expertise, research, and insights have made this book a comprehensive and valuable resource. Each chapter reflects the dedication and passion of the authors who have worked tirelessly to highlight the multifaceted potential of millets in addressing some of the most pressing issues of our time. The diversity of perspectives from Brazil, India, and beyond underscores the importance of interdisciplinary collaboration in tackling global challenges.

A special note of appreciation goes to:

- Beatriz Lucia Salvador Bizotto, Dr. Maria Emilia Camargo, and Dr. C.B. Singh, whose insightful work on the ecological sustainability of corn production in Brazil provides valuable lessons for sustainable agricultural practices.
- Dr. Noopur Gautam, Chaudhary Mohd Kashim, and Dr. Shikha Singh, who have expertly highlighted the nutritional and economic value of millets as a critical resource for food security.
- Shubhangi Nigam and Dr. Dinesh Kumar Gupta, whose analyses of millet agriculture's role in improving rural livelihoods and achieving SDG targets are inspiring and impactful.
- Dr. Krishnan Kutty for his in-depth study of pearl millet production trends and its implications for the agricultural landscape in India.
- Prof. (Dr.) Anil S. Maheshwari for his exploration of millet's role in Maharashtra and its influence on food security in the region.
- Prof. Debesh Bhowmik and Dr. Jyoti Kumari Mishra for her detailed work on the economic viability of millet cultivation and its contribution to sustainable livelihoods.

- Dr. R.M. Chitnis and Dr. Vijay Kulkarni for their exploration of the health benefits of millet-based products, which are crucial for promoting immunity and resilience in communities.

We also extend our heartfelt thanks to Springer Nature for their unwavering support in publishing this volume and to all the editors, researchers, and colleagues who have contributed to shaping the narrative of this book. We would also like to acknowledge Chaudhary Mohd Kashim, Niraj Sharma, and Priyal Chaturvedi for their continuous support during the editing phases of this book.

Lastly, I would like to express my deepest appreciation to my co-authors, editors, and the entire team for their collaboration, patience, and commitment to bringing this important work to fruition. Together, we have laid the foundation for further exploration, policy advocacy, and innovation in the promotion of millets as a cornerstone for sustainable agriculture, human health, and environmental stewardship.

Let this book serve as a call to action for a future where millets play a central role in building resilient food systems and promoting sustainability on a global scale.

C. B. Singh
Ankita Jasmine Lall
Shilpa Mishra

Contents

Contributors

Angshumitra Banerjee The PSBB Millennium School Gerugambakkam, Chennai, Tamil Nadu, India

Debesh Bhowmik Faculty-Business and Accountancy, Lincoln University College, Petaling Jaya, Selangor, Malaysia

Beatriz Lucia Salvador Bizotto Department of Social and Applied Sciences, UNIFACVEST University Center, Lages, SC, Brazil

Global Edu Leaders Forum, Varanasi, Uttar Pradesh, India

Maria Emilia Camargo Mestrado Acadêmico em Educação Profissional e Tecnológica—CTISM/UFSM, Federal University of Santa Maria, Santa Maria, RS, Brazil

CNPQ Researcher, Brasilia, Brazil

R. M. Chitnis MIT World Peace University, Pune, Maharashtra, India

Uzma Fareen Community Institute of Commerce & Management Studies, Bangalore, Karnataka, India

Noopur Gautam Institute of Food Technology, Bundelkhand University, Jhansi, Uttar Pradesh, India

Dinesh Kumar Gupta Department of Economics, Rajkiya Mahavidyalaya Amori, Champawat, Uttarakhand, India

Chaudhary Mohd Kashim Institute of Food Technology, Bundelkhand University, Jhansi, Uttar Pradesh, India

Jaspreet Kaur Department of Food Technology, Institute of Engineering and Technology, Bundelkhand University, Jhansi, Uttar Pradesh, India

Vijay Kulkarni Director Corporate & Executive Program, Ajeenkya DY Patil University, Pune, Maharashtra, India

V. Krishnan Kutty Government College Kodanchery, University of Calicut, Kozhikode, Kerala, India

Ankita Jasmine Lall Department of Economics, Christ Church College, Kanpur, Uttar Pradesh, India

Anil S. Maheshwari School of Engineering, Ajeenkya DY Patil University, Pune, Maharashtra, India

Khushboo Mansoori Department of Food Technology, Institute of Engineering and Technology, Bundelkhand University, Jhansi, Uttar Pradesh, India

Jyoti Kumari Mishra Department of Banking Economics and Finance, Bundelkhand University, Jhansi, Uttar Pradesh, India

Shilpa Mishra Institute of Economics and Finance, Bundelkhand University, Jhansi, Uttar Pradesh, India

Shubhangi Nigam Department of Food Technology, Institute of Engineering and Technology, Bundelkhand University, Jhansi, Uttar Pradesh, India

Sushila Rajan Community Institute of Commerce & Management Studies, Bangalore, Karnataka, India

Sachin Rajas Academic Planning, Ajeenkya DY Patil University, Pune, Maharashtra, India

Nandini Saini Department of Food Technology, Institute of Engineering and Technology, Bundelkhand University, Jhansi, Uttar Pradesh, India

Mithileysh Sathiyanarayanan Research & Development, MIT Square, London, Bengaluru, India

Anushka Sharma Department of Food Technology, Institute of Engineering and Technology, Bundelkhand University, Jhansi, Uttar Pradesh, India

C. B. Singh Institute of Economics and Finance, Bundelkhand University, Jhansi, Uttar Pradesh, India

Shikha Singh School of Management, Ajeenkya DY Patil University, Pune, Maharashtra, India

Manas Vijay Department of Food Technology, Institute of Engineering and Technology, Bundelkhand University, Jhansi, Uttar Pradesh, India

Priyanshu PalManas Vijay Department of Food Technology, Institute of Engineering and Technology, Bundelkhand University, Jhansi, Uttar Pradesh, India

Abbreviations

ACC	Agro-climatic conditions
ANC	Anti-nutritional components
ANVISA	Agência Nacional de Vigilância Sanitária
APEDA	Agricultural and Processed Food Products Export Development Authority
ARDL	Autoregressive distributed lag
BBA	Bibliometric-based analysis
BBS	Bachelor of business studies
BM	Barnyard millet
BMI	Body mass index
BTM	Browntop millet
C3 plant	3-phosphoglyceric acid
CACP	Commission for Agricultural Costs and Prices
CAGR	Compound annual growth rate
CIAE	Central Institute of Agricultural Engineering
CV	Coefficient of variation
DASH	Dietary approaches to stop hypertension
DNA	Deoxyribonucleic acid
ECO	Ecological
ERAS	Environmentally relevant activities
ES	Ecological sustainability
FAO	Food and Agriculture Organization
FPO	Farmer Producer Organization
GABA	Gama-amino butyric acid
GFS	Global food system
GHG	Greenhouse gas reduction
GIS	Geographic information system
GM	Green management
GoI	Government of India
GR	Green Revolution
HDL	High-density lipoprotein
HDL-C	High-density lipoprotein cholesterol
HLP	Home level processing
HYV	High yielding variety

ICAR	Indian Council of Agricultural Research
ICRISAT	International Crops Research Institute for the Semi-Arid Tropics
IPCC	Intergovernmental Panel on Climate Change
IPM	Integrated pest management
ITC	Input tax credit
IYOM	International Year of Millets
JELS	Journal of Environmental Law and Socio-Environmentalism
KM	Kodo millet
LDL	Low-density lipoprotein
LDL-C	Low-density lipoprotein cholesterol
MPS	Marginal propensity to save
MSPs	Minimum support prices
NFSM	National Food Security Mission
NGO	Non-governmental organization
NNHW	Nourishment needed for human wellness
NSSO	National Sample Survey Office
NUS	Neglected and underutilized species
OS	Organic systems
OWS	Other welfare schemes
PDM	Political decision makers
PDS	Public distribution system
PM POSHAN	Pradhan Mantri Poshan Shakti Nirman
PUFA	Polyunsaturated fatty acids
R&D	Research and development
RTEFP	Ready-to-eat food products
SA	Sustainable agriculture
SAPs	Special agriculture products
SCECB	Sustainable corn ecological and commercial balance
SDG 12	Sustainable Development Goal 12
SDG 2	Sustainable Development Goal 2
SDG 3	Sustainable Development Goal 3
SDGs	Sustainable Development Goals
SHGs	Self-help groups
SHRFC	Soaking, heating, roasting, fermenting, cooking
SoI	Statement of intent
SPSS	Statistical Package for the Social Sciences
SUS or SUST	Sustainability
TC	Total cholesterol
UEA	University of the State of Amazonas
UFBA	Federal University of Bahia
UFCG	Federal University of Campina Grande
UFFS	Federal University of the Southern Border Erechim Campus
UFPB	Federal University of Paraíba
UFRA	Federal Rural University of Amazon
UFRGS	Federal University of Rio Grande do Sul

UFRJ	Federal University of Rio de Janeiro
UFRN	Federal University of Rio Grande do Norte
UFS	Federal University of Sergipe
UFSCar	Federal University of São Carlos Center of Agrarian Sciences
UFU	Federal University of Uberlândia, Institute of Agrarian Sciences
UnB	University of Brasília
UNGA	United Nations General Assembly
UNSP	University State Paulista
USD	United States Dollar
USDA	United States Department of Agriculture
USP	University of São Paulo
UTFPR	Federal Technological University of Paraná
VAP	Value-added products
VECM	Vector error correction model
VLDL-C	Very-low-density lipoprotein cholesterol
WBNP	Wheat Based Nutrition Programme
WFP	United Nations World Food Programme

Evaluating the Ecological Sustainability of Corn Production In Brazil

1

Beatriz Lucia Salvador Bizotto, Maria Emilia Camargo, and C. B. Singh

Abstract

Corn is one of the most important basic cultures in Brazil, with a long history of cultivation and significant economic and social impact. The expansion of corn production in Brazil has been driven by the growing demand for food, rations and biofuels, both nationally and internationally. However, this growth also brought significant environmental challenges such as soil erosion, water pollution and biodiversity loss. Water scarcity and pollution are also major concerns in corn production. To reduce water use and pollution, farmers can adopt precision irrigation techniques and reduce the use of chemical fertilizers and pesticides. Biological agriculture is another sustainable practice that can help reduce the environmental impact of corn production. In this context, we have the following question: What are the future perspectives for ecological sustainability in corn production? To answer this research question, we conducted a bibliometric research to identify the studies being conducted in Brazil. The methodologies used are qualitative and quantitative with a bibliometry on the Google Academic platform. The research period was from 2019 to 2023. The keywords used were: "Agricultural Development", "Ecological Sustainability", "Corn Production". Sixty documents were located. This study pointed out to the relevance of the

B. L. S. Bizotto
UNIFACVEST University Center, Lages, SC, Brazil

Global Edu Leaders Forum, Varanasi, Uttar Pradesh, India

M. E. Camargo (✉)
Federal University of Santa Maria, Santa Maria, RS, Brazil

CNPQ Researcher, Brasilia, Brazil

C. B. Singh
Universidade Bundelkhand Jhansi, Jhansi, Uttar Pradesh, India
e-mail: dr.cbsingh195@bujhansi.ac.in

C. B. Singh et al. (eds.), *Global Millets Production for a Sustainable Future*,
https://doi.org/10.1007/978-3-031-88157-2_1

theme sustainability today, especially with regard to agribusiness. Another international research is suggested to compare the studies of this research.

Keywords

Agricultural development · Corn production · Sustainability · Ecological

1.1 Introduction

Agriculture has long been a significant part of Brazil's economy, with the country being one of the world's largest producers and exporters of agricultural products. Over the years, the agricultural sector of Brazil has undergone a significant transformation of a system based mainly on subsistence to a modern and export -oriented industry, as well as challenges in meeting international standards.

Ecological challenges in corn production in Brazil are significant, but there are also many sustainable practices that can help mitigate these challenges. One of the most pressing issues is soil erosion and degradation, which can be caused by monoculture and intensive crop. To solve this problem, farmers can implement conservation crop, which involves minimal soil disturbance and the use of coverage crops to protect the soil from erosion. Culture rotation is also an effective practice to reduce soil erosion and improve soil fertility.

Future prospects for ecological sustainability in corn production in Brazil are promising, but there is still a lot of work to be done. Advances in technology and research are making it possible to develop new sustainable agricultural practices and improve existing ones. Collaboration between stakeholders, such as farmers, researchers, political decision makers and consumers, is also essential to promote sustainable agriculture. Consumers can play a significant role in promoting sustainable agriculture, demanding and supporting products produced in an environmentally correct and socially responsible way. Working together, we can guarantee that corn production in Brazil is sustainable, economically viable and socially responsible.

1.2 Theoretical Reference

1.2.1 Agricultural Development

Agriculture has been the backbone of human civilization since the beginning of time. He played a crucial role in supplying food, fiber and other essentials for human survival. With the increase of the global population, the search for food and other agricultural products has also increased. This led to the development of new technologies and techniques to improve productivity and agricultural efficiency. However, the surprising growth of agricultural productivity still faces other serious obstacles to its expanded reproduction from the socioeconomic and environmental point of view (Gonçalves, 2004; Barbosa & Queluz, 2021).

Imposition to urban -industrial development occurred during the nineteenth century, everywhere, a great hesitation between the promotion of organized agriculture according to the manufacturing model—with a clear separation between the directive and executive functions of the production process—and support to the adaptation of peasant agriculture to the challenges it faces. In France of the Second Empire, when the commercial treaty with Britain was established, there was a large wave of government and intellectual compliments in favor of the industrialization of agricultural production (Veiga, 2023).

Perhaps it is possible, then, to talk about a kind of state intervention standard aiming at the modernization of capitalist agriculture. The two basic lines of these policies would be the following: public investments in investigation, extension, infrastructure and credit encourage supply growth at a faster pace than demand growth; The trade policy generates a stable drop in food prices, compatible with the maintenance of current remuneration acceptable for the farmer‘s work and also compatible with the regulation of the rural exodus required, both by the urban labor market and the need Gradually increase the size of production units. Of course, this pattern did not prevent the different treatment of the various branches of the agricultural sector, generating different rhythms of technological progress (Veiga, 2023).

Corn has been one of the viable alternatives and the possibility of combining different agroenergetic biomass as a renewable source of energy for the production process. The search for sustainability and the growing energy demand has directed humanity's attention to biofuels. In this sense, corn has drawn attention as a raw material for ethanol production in Brazil (Douradinho, 2023)

1.2.2 Corn Production

The importance of agriculture in corn production as an alternative for new jobs and income for families and food industry cannot be exaggerated. Corn is one of the most cultivated cultures in the world and is an important basic food in many diets. It is used in many ways, from animal feed to human consumption and even in the production of biofuels. In fact, corn is so versatile that it can be found in almost every corner of the world. One of the most important factors that make corn production significant is its nutritional value. Corn is a good source of carbohydrates, fiber, vitamins and minerals.

Consumption is expected to fall lower (−1.6%), higher than production in this crop (1.161 × 1.150 billion tons), but reversing the next, when production should exceed consumption by 27.2 million tons (USDA, 2023). It is expected that United States, China and Brazil will produce 65.9% of the 1.15 billion tons of the current crop (2022/23), a portion that should be changed in the first forecast for the next crop (2023/24) of Department of US Agriculture (USDA), which should be 65.1%. Production, consumption, exports, imports and final stocks should increase for the next harvest (CONAB, 2023; Coêlho, 2023).

Brazil is the largest exporter of corn and expects significant growth in production, with increased area and productivity. The reduction of rainfall and dry weather speeds up the harvest, and the internal and external demands remain heated, especially with the greatest interest of China in Brazilian corn. The largest Brazilian producers are Mato Grosso, Paraná, Mato Grosso do Sul, Goiás and Minas Gerais. Corn cultivation allows three annual crops, the second being the highest production. Corn has also been used in ethanol production, expected to increase 42% by 2023/24 (Coêlho, 2023).

1.2.3 Ecological Sustainability

Ecological sustainability with environmental and social practices induces development. Income generation through ecological sustainability. One way to achieve ecological sustainability is through sustainable social and environmental practices. In today's world, ecological sustainability has become a fundamental issue that attracted people's attention around the world. The concept of sustainable development has emerged as a tool for facing the environmental, social and economic challenges we face today. The concept of sustainability is not just about protecting the environment; It also covers social and economic aspects that are equally important. Ecological sustainability is the use of natural resources in a way that perpetuate (Rocha, 2016).

In general, the three objectives that guide the ecological economy are: ecological sustainability, social distribution and efficient use of resources, in a new sustainable economic development paradigm that is opposed to looking for unlimited economic growth. Moreover, among the methods and instruments aimed by ecological economists, are the study of the application of environmental public policies, based on the precautionary principle, the development of new indicators and sustainability indices in the economy, the ecological macroeconomics, the evaluation of the capacity of capacity of Support - the population or material limit - of ecosystems, the discussion about valuation and recurring externalities about environmental services, and their possible incommensurability (Medeiros, 2023).

A theoretical divergence is present among ecological economists and indicates not only the incentive to plurality, but also the non -rigidity of a revolution in the economic paradigm that allows its constant improvement. The ways to meet ECO's objectives may be different, but these goals, constant criticism of the current production model and the physical limit belief for production that maintain the theoretical convergence of scientists in the field (Medeiros 2023; Albuquerque & Lima, 2023).

The term "sustainability" is quite usual today and its very recent concept is implied in a process of resistance and awareness of the risks caused by the excessive exploitation of nature and the intrinsic irresponsibility to technological and industrial culture. However, with the rise of the very environmental in the late twentieth century, the concept of sustainability gained impetus. Sustainability refers to the ability to meet our present needs without compromising the ability of future generations (Andrade, 2023).

1.3 Methodology

The method used was bibliometric based on observations of the number of scientific publications in 2019 to 2023. To conduct this research, Google Academic was used. The keywords used were "agricultural development", "ecological sustainability""corn production". Seeking to explain from a literary point of view, projects, research articles that address adoptions in Brazil, with 60 results: types of resources, authors, year and title. Access occurred between January 3 and 4, 2024 (Deslauriers & Kérisit, 2023; Vergara & Peci, 2003; Silva & Vergara, 2015; Yin, 2005).

1.4 Results

Sustainable agriculture is becoming increasingly important as the global population continues to grow and natural resources become more limited. In particular, corn production, a basic culture for many countries, has been the target of scrutiny due to its environmental impact. In this article, we will conduct a bibliometric analysis of the latest research on sustainable agriculture and corn production of 2019–2023. We will examine the arguments that prioritize ecological sustainability, improved corn production and sustainable practices over short -term profits. We will also analyze counter-compliances that challenge these perspectives and explore the complex relationship between sustainable agriculture and economic development.

The methodology used is qualitative and quantitative with the realization of a bibliometry in Google Academic and Scopes. The research was conducted only with Brazilian articles. The period comprised is from 2019 to 2023. The keywords used were "agricultural development", "ecological sustainability"and "corn production". 60 documents were located. After reading were excluded articles from other Portuguese or Spanish-speaking countries, books and reports. After exclusion only 37 documents were and we have the results in the following tables. Following is the Table 1.1 we have the list of all documents used in the study.

In Table 1.1 we have the 37 authors listed in the studies, name of the authors, work title, year of publication, journal or university, citations received, type of document if it was doctoral dissertation, master's dissertation, completion work or academic article. It was observed that no author published more than once with the same approach.

In Table 1.2 we have the productions by universities or periodicals, which published more than one document.

It is observed that USP obtained two publications a doctoral dissertation and a master's dissertation. The Federal University of Rio de Janeiro obtained a doctoral dissertation and two master's dissertations. The Federal University of Pelotas two master's dissertations. Ministry of Education Federal University of Sergipe a doctoral dissertation, a master's dissertation. Federal University of Paraíba Two Course Conclusion Papers (TCC). Federal University of the Southern Frontier A Master

Table 1.1 List of documents used in the study

Author	Title	Year	University	Citations	Type/ document
Vilanova, N. R.	Evaluation of growth and degree of herbivory in corn plants cultivated with borders of guandu beans, crotalaria and napier grass	2019	University of Brasília		Course Conclusion Work
Vernini, A. A.	Embodied energy in the production of biodiesel from crambe crops	2019	UNSP—University State Paulista		Thesis
Lopes, E. V., & Padilha, N. S.	Setbacks in the risk communication system for pesticide labeling: Anvisa's classification	2019	Journal of Environmental Law and Socio-Environmentalism	05	Article
Morais, P. D. P.	Landscape fragmentation and ecological corridors in the MATOPIBA region	2019	University of Brazilia	01	Course Conclusion Work Faculdade de tecnologia
Lopes, J. S.	Green oil in Bahia	2019	Federal University of Bahia		Course Conclusion Work Faculdade de comunicação
Silva, J. R. A. D.	Non-transgenic corn—Characterization and product formulation	2019	Federal University of Paraíba		Course Conclusion Work
Silva, F. F. D.	Organization of social control in the agroecological transition of the Pirituba settlement	2019	Federal University of Campina Grande		Dissertation
Lima, R. S. D.	Perception of water and place in the sub-basin of the River Lajes/ Sergipe	2019a	Federal University of São Carlos Center of Agrarian Sciences		Thesis
Mendonça, M. V.	Ecological corridor between the protected areas of Carajás and Terra do Meio, Pará	2019	Federal University of Sergipe	1	Dissertation
Destefani, W. C.	An ethnomathematical study with family members and first-grade students at an agroecological school in the municipality of Águia Branca-ES	2019	Federal University of Tocantins	4	Dissertation

(continued)

Table 1.1 (continued)

Author	Title	Year	University	Citations	Type/ document
Quevedo, M. G.	Estimating the economic gains from rotating rice with soybeans in the Lowlands	2020	Federal Rural University of Rio de Janeiro		Dissertation
Neto, B. S.	Relations between added value and wealth in the analysis-diagnosis of agricultural systems	2020		2	Article
Santos, L. S. C., & Garcia, R. P. M.	Agroecology: a strategy to fight for the strengthening and resistance of peasant youth	2020	Kiri-Kerê-Pesquisa em Ensino	1	Article
Aguilar López, F. M.	Identifiable changes in the level of Hemerobia resulting from the dynamics of land use in the Rio Grande Comitán Hydrographic Basin (Chiapas, Mexico). an analysis of the period 1999–2019	2020	Repositorio Institucional UFSCar		Thesis
Torres, C. J. F.	Methodological bases for inserting the water-energy-agriculture nexus concept into intersectoral planning and management models	2020	Universidade Federal da Bahia	01	Thesis
Silva, D. I. D. C.	Meanings attributed by family farmers to the practice of polyculture in its relationship with health and the environment	2020	Universidade Federal da Bahia Bahia		Dissertation
Faria, V. L. D.	Soil quality under banana cultivation in organic and conventional production systems in Gonçalves	2020	Universidade Federal de Itajubá	03	Dissertation
Drehmer, A. C. B.	Agroecological family farming systems - first District of Canguçu/RS: a multidimensional analysis for rural development	2020	Universidade Federal de Pelotas		Dissertation

(continued)

Table 1.1 (continued)

Author	Title	Year	University	Citations	Type/ document
Reinach, M. H. D. M.	Controversies over the Social Dimension of Agroecology: A comparison between the political discourses of the FAO and La Via Campesina	2020	Universidade Federal Rural do Rio de Janeiro		Dissertation
Andrade, A. S. D.	Information for public development policies in Rondônia	2020	USP - Universidade de São Paulo		Thesis
Cargnelutti, D., Bampi, E., de Melo Santiago, G., da Luz, V. C., Garbin, E., Castamann, A., & Mossi, A. J.	Emerging technological solutions for sustainable agriculture: efficient microorganisms. García, LMH Agroecology: ecological principles and foundations applied in the search for sustainable production, Canoas	2021	Mérida Publishers		Article
Brandão, T. F. B.	Women's association "Resgatando sua História": social organization and management in rural northeastern Brazil, Porto da Folha, Sergipe	2021	State University of Campinas		Thesis
Pieniz, F. G.	Added value and sustainable development: the case of rural agro-industries in the municipality of Cerro Largo	2021	Federal University of The Southern Border Cerro Largo Campus		Dissertation
Ferreira, C. O.	The agri-environmental issue in the TVT thematic axis: a case study at the Nova Itamarati State School, Itamarati II Settlement - Ponta Porã, MS	2021	Federal University Of Grande Dourados		Dissertation
Batista, C. L. R., Stoffel, J.	Agroecology and organic production: characteristics that distinguish and/or bring sustainable production systems closer together	2022	COLÓQUIO—Journal of Regional Development—Faccat	02	Article

(continued)

Table 1.1 (continued)

Author	Title	Year	University	Citations	Type/ document
Melo, R. L. D. M.	Agroforestry systems and the legacy of João Ambrósio de Araújo Filho: Subsidies for the creation of the appropriate livestock institute in the municipality of Gurjão-Paraíba	2022	Federal University of Paraíba		Course Conclusion Work
Vallejo, L. G.	Agroecological systems: information regime of a university campus as a Living Lab. 2022	2022	Federal University of Rio de Janeiro—Ufrj		Thesis
Sorrini, T. B.	Restoring the campus: involving students and planting trees for the environmental transformation of ESALQ/USP	2022	USP		Dissertation
Santos, J. F. S.	Selection of corn genotypes and physiological responses to water deficit in a semi-arid region	2023	Federal University of Sergipe		Dissertation
SILVA, R. D. O.	Socio-economic and agro-environmental analysis of family farming in the rural community of Fazendinha, Magalhães Barata, Pará	2023	Federal Rural University of Amazon		Course Conclusion Work
Mezalira, S. M.	From the limiting situation of agrotoxins to the unprecedented and feasible expressed by agroecology: understandings and practices of teachers and students from rural schools	2023	Federal University of Rio Grande do Sul		Thesis
Brito, A. C. L.	The fundamental right to adequate food of indigenous peoples in Brazil and the cultivation of transgenic foods in their territories	2023	Universidade do Estado do Amazonas—UEA		Dissertation

(continued)

Table 1.1 (continued)

Author	Title	Year	University	Citations	Type/ document
Markovski, A. M.	The evolution of agroecology in the municipality of Itatiba do Sul/RS	2023	Federal University of The Southern Border Erechim Campus		Course Conclusion Work
Bortonio, N. D. S.	Association of Bacillus sp. with organomineral 06.30. 00 in corn cultivation	2023	Federal University of Uberlândia Institute of Agrarian Sciences		Course Conclusion Work
Lima, P. C. D.	New rural in the immediate region of Patos-PB: the construction of a textbook for the seventh year of elementary school geography	2023	Federal University of Rio Grande do Norte		Dissertation
Reginatto, M.	Organic no-till system: crop rotations and mechanical control of spontaneous plants in southwest Paraná	2023	Federal Technological University of Paraná		Thesis

Source: Authors 2023

Dissertation and Course Conclusion Paper. And the Federal University of Bahia a doctoral dissertation, a master's dissertation and a course completion work.

The next result is from Table 1.3 where we have the amount of productions per year of the period studied.

Firstly, it was 2019 where it had 11 publications, followed by 2020 with 10 publications, third was 2023 with 08 publications and 2021 and 2022 obtained 04 annual publications.

On Table 1.4 we have the authors who were cited at least twice in the studied period.

The author Lopes and Padilha (2019), Setbacks in the System of Risk Communication in Pesticide Labeling: Anvisa's Classification, had 5 citations. LIMA (2019a, 2019b), Potential of Clove Essential Oil on Phytopathenic Fungi in Corn Seeds, also had 05 quotes. Destefani (2019), obtained 04 citations, an ethnomathhematical Study With Family Members and Students in the First Year of Elementary School at An Agroecological Schoool in the Municipal Eagle of White-ES. The other authors had 03, 02 or one quote only. In Table 1.5 we have the objectives of the documents that had at least two citations.

In this Table 1.5 it is possible to observe what was the objective of the study, thus it is believed that facilitates researchers interested in continuing research in the study studied as well as the results found. In short, we have to: regulatory setbacks in the pesticide control system directly and immediately affect rural workers, who are not always trained for the management of pesticides and are excessively exposed without proper communication and information on risks to Your health. The

Table 1.2 Number of productions by university

University	Thesis/ qt	Dissertation/ qt	Course conclusion work	Total
USP	1	1	0	2
Universidade Federal Rural do Rio De Janeiro	1	2	0	3
Universidade Federal de Pelotas	0	2	0	2
Ministério da Educação Universidade Federal De Sergipe	1	1	0	2
Universidade Federal da Paraíba	0	0	2	2
Universidade Federal da Fronteira Sul		1	1	2
Universidade Federal da Bahia	1	1	1	3

Source: Authors 2023

Table 1.3 Years and quantity of production year

Production	Quantity
2019	11
2020	10
2021	4
2022	4
2023	8
Total	**37**

Source: Authors 2023

chemical indicators had chemical conditions favorable to the development of corn in organic systems, and especially in conventional systems, where, in addition to the deposited organic material, chemical inputs were added. The results observed in this study confirm the antifungal capacity of India's clove essential oil.

Thus it is possible to infer that agricultural development should give priority to ecological sustainability. Ecological sustainability guarantees the viability of long -term agriculture, improving harvest yield and reducing environmental damage. By using sustainable practices, such as crop rotation, natural pest control and minimal mobilization, farmers can improve soil health, which in turn leads to better crop yields. In addition, sustainable practices can reduce the amount of chemicals and pesticides used, which can damage the environment and human health. By giving priority to ecological sustainability, farmers can create a more resilient and better equipped agricultural system to resist climate change and other environmental challenges.

Ecological sustainability should not occur at the expense of economic growth. Agricultural development can support economic growth, providing employment opportunities, increasing food safety and supporting rural economies. Some practices that improve ecological sustainability can be too expensive for farmers, especially those of developing countries. In addition, ecological sustainability can be difficult to achieve in certain areas or climates, which can make farmers difficult to adopt sustainable practices. Although ecological sustainability is important, it should not be pursued at the expense of economic growth and the means of subsistence of farmers.

Table 1.4 Relationship of the authors and works that were cited

Author	Title	Year	University	Citations	Doc
Lopes, E. V., & Padilha, N. S.	Setbacks in the system of risk communication in pesticide labeling: Anvisa's classification	2019	Revista de Direito Ambiental e Socioambientalismo	05	Article
Lima, T. S.	Potential of clove essential oil on phytopathogenic fungi in corn seeds	2019b	Federal University of Campina Grande	05	Dissertation
Destefani, W. C.	An ethnomathematical study with family members and students in the first year of elementary school at an agroecological school in the municipality of Águia Branca-ES	2019	Federal Rural University of Rio de Janeiro	04	Dissertation
Faria, V. L. D.	Soil quality under banana cultivation in organic and conventional production systems in Gonçalves	2020	Federal University of Itajubá	03	Dissertation
Batista, C. L. R., & Stoffel, J.	Agroecology and organic production: characteristics that distinguish and/or bring sustainable production systems closer together	2022	COLLOQUIUM—Journal of Regional Development—Faccat	02	Article
Neto, B. S.	Relations between added value and wealth in the analysis-diagnosis of agricultural systems	2020	Rural Extension	2	Article
Torres, C. J. F.	Methodological bases for inserting the water–energy–agriculture nexus concept into intersectoral planning and management models	2020	Federal University of Bahia	01	Thesis
Santos, L. S. C., & Garcia, R. P. M.	Methodological bases for incorporating the water-energy-agriculture nexus concept into intersectoral planning and management models	2020	Kiri-Kerê-Research in Teaching	01	Article
Mendonça, M. V.	Ecological corridor between the protected areas of Carajás and Terra do Meio, Pará	2019	Federal University of Tocantins	01	Dissertation
Morais, P. D. P.	Landscape fragmentation and ecological corridors in the MATOPIBA region	2019	University of Brasília	01	**Course Conclusion Work**

Source: Authors (2023)

Corn production should be improved to satisfy the growing demand. Corn is an important component of many food products and their production can support rural economies. Improving corn production can lead to lower prices for consumers, which can increase access to food and improve food safety. In addition, improving corn production can reduce the environmental impact of corn cultivation, reducing the amount of land needed to produce the same amount of corn.

Corn production should not be the only focus of agricultural development. Culture diversification can improve soil health and reduce the risk of crop failure due to diseases or pests. Focusing only on corn production can lead to monoculture, which can reduce biodiversity and make the agricultural system more vulnerable to environmental challenges. Although corn is an important culture, it should not be produced at the expense of other cultures and long -term health of the agricultural system.

Agricultural development should give priority to sustainable practices over short -term profits. Sustainable practices can improve soil health, water quality and biodiversity. By using sustainable practices, farmers can create a more resilient agricultural system that can resist environmental challenges. In addition, sustainable practices can help reduce the environmental impact of agriculture, which can lead to long -term benefits for both farmers and the environment.

Short-term profits are necessary for farmers to invest in sustainable practices. Farmers need to make a profit to invest in sustainable practices such as crop rotation, natural pest control and minimal crop. Short -term profits can help farmers face economic challenges and invest in the health of their long -term explorations. Although sustainable practices are important, farmers need to be able to make a living to adopt these practices and create a more sustainable agricultural system.

In conclusion, sustainable agriculture and corn production are complex issues that require careful balance between ecological sustainability, economic growth and short -term profits (Garcia, 2021). Although giving priority to ecological sustainability is important for long-term health of the agricultural system and the environment, it should not be continued at the expense of economic growth and farmers'subsistence means. Similarly, although corn production is an important component of many food products, it should not be produced at the expense of other cultures and long -term health of the agricultural system. Ultimately, achieving sustainable agriculture will require a combination of sustainable practices, economic growth and careful management of natural resources.

1.5 Final considerations

This study aimed to be the future perspectives for ecological sustainability in the production of corn in Brazil. To answer the objective we have as a question of research which studies are being carried out in Brazil? After the bibliometry, it was realized that there was a limited number of studies with the key words of the search. Another factor that also caught the researchers' attention is that the authors no longer published research with the same subjects that were addressed in the doctoral

Table 1.5 Publications with at least 02 citations

Author	Title	Objectives	Results
Lopes, E. V., & Padilha, N. S. (2019)	Setbacks in the system of risk communication in pesticide labeling: Anvisa's classification	Analyze the toxicological reclassification on the labeling of agrochemical products in Brazil, as determined by ANVISA (2019), contextualizing the topicality of the issue with the principle of Information and parameters granted by the Escazú Agreement, the Aarhus Convention and the Sustainable Development Goals (SDGs) of the United Nations Agenda 2030, referring to health, well-being, consumption and sustainable production	"The regulatory setbacks in the pesticide control system directly and immediately affect rural workers, who are not always trained to handle pesticides and are overexposed without proper communication and information about the risks to their health. In the recent toxicological delimitation of pesticides by Anvisa, on July 23, 2019, the choice of parameters for the toxicological reclassification of pesticides did not favor clear and objective information to the user and consumer, since the new parameter used by Anvisa does not favor precaution and risk communication, which is evident from the excessive increase in the number of authorized products. And, as much as it can be said that parameters similar to international regulations are being applied, there is a clear deficit of risk security in Brazil, where pesticides that have already been banned in the rest of the world are being released"
Lima, T. S (2019b)	Potential of clove essential oil on phytopathogenic fungi in corn seeds	The aim of this study was to evaluate the inhibitory potential of clove essential oil (Syzygium aromaticum) on phytopathogenic fungi in vitro and on maize seeds	"Their results showed that the essential oil has a fungistatic and fungicidal effect on the mycelial growth and germination of conidia of Fusarium sp. They also report that the effect is greater as the concentration of the oil increases. The results observed in this study confirm the antifungal capacity of clove essential oil. In the in vitro test, the oil completely inhibited the growth of the fungi evaluated even at low concentrations, while in the in vivo test, it reduced the incidence of phytopathogens in corn seeds"

(continued)

Table 1.5 (continued)

Author	Title	Objectives	Results
Destefani, W. C. (2019)	An ethnomathematical study with family members and first-grade students at an agroecological school in the municipality of Águia Branca-ES	To investigate the generation and dissemination of mathematical knowledge from an ethnomathematical perspective allied to the principles of agroecology. We proposed pedagogical actions to work on mathematics content based on the relationships that exist with agroecology in the students' daily lives, valuing their culture and the activities of the countryside, reflecting the collaborative work of the school and families, bringing favorable daily situations that are offered by rural education to ethnomathematical experiences	"By establishing how Ethnomathematics and Agroecology relate to each other based on the relationships that exist in the daily lives of the students surveyed, we were able to propose pedagogical actions to work on mathematics content based on these relationships. We carried out an investigative activity of a qualitative nature. The development of the research was based on the relationships between the observations, workshops, conversation circles, reports, interviews, photographic records provided by the students and their families, on the rural properties, in their daily activities, as well as in the pedagogical spaces of the school"
Faria, V. L. D. (2020)	Soil quality under banana cultivation in organic and conventional production systems in Gonçalves	To evaluate the quality of the soil in commercial banana plantations under organic and conventional production systems, by analyzing its main physical, chemical, microbiological and visual attributes	"The results indicate that the cultural practices and soil management adopted in the organic and conventional production systems contributed to a significant contribution of organic matter attributed to the constant deposition of banana biomass, with no difference to the reference area of native forest, thus maintaining the physical, microbiological and visual quality of the soil. The chemical indicators showed favorable chemical conditions for the development of the plantation in the organic systems, and especially in the conventional systems, where, in addition to the organic material deposited, chemical inputs were added"

Source: Authors (2023)

theses, master's dissertation and also course completion work. This with certain is an issue that deserves attention. And generated questions like why did the authors not publish anymore about the topic?

This study pointed out to the relevance of the theme sustainability today, especially with regard to agribusiness. However, in addition to development and implementation, it is necessary to stick to ecological sustainability issues with the implementation of public policies to mitigate the causing effects of agrochemical products and deforestation. Agribusiness is known to involve production for food and commercial purposes. In this respect, the concept of sustainability where the economic, ambient and social is involved.

However, with the rise of environmentalism in the late twentieth century, the concept of sustainability gained impulse. Sustainability refers to the ability to meet our present needs without compromising the ability of future generations to meet their own needs. It covers a wide variety of practices, from the reduction of carbon emissions and the conservation of water and energy to the use of sustainable materials and the promotion of social equity. Environmentalism brought a change in the way we view progress and development. The need for sustainable practices to ensure that our planet can continue to support life is stressed. Sustainability has become a crucial consideration in all aspects of our lives, from the food we consume to the products we use. The concept of sustainability has been extensively studied and researched, and guidelines and policies have been developed to promote sustainable practices.

Agricultural development should give priority to sustainable practices over short -term profits. Sustainable practices can improve soil health, water quality and biodiversity. By using sustainable practices, farmers can create a more resilient agricultural system that can resist environmental challenges. In addition, sustainable practices can help reduce the environmental impact of agriculture, which can lead to long-term benefits for both farmers and the environment.

Short -term profits are necessary for farmers to invest in sustainable practices. Farmers need to make a profit to invest in sustainable practices such as crop rotation, natural pest control and minimal crop. Short -term profits can help farmers face economic challenges and invest in the health of their long -term explorations. Although sustainable practices are important, farmers need to be able to make a living to adopt these practices and create a more sustainable agricultural system.

In conclusion, sustainable agriculture and corn production are complex issues that require careful balance between ecological sustainability, economic growth and short -term profits. Although giving priority to ecological sustainability is important for long -term health of the agricultural system and the environment, it should not be continued at the expense of economic growth and farmers'subsistence means. Similarly, although corn production is an important component of many food products, it should not be produced at the expense of other cultures and long -term health of the agricultural system. Ultimately, achieving sustainable agriculture will require a combination of sustainable practices, economic growth and careful management of natural resources. And finally, another international research is suggested to compare the studies of this research.

References

Aguilar López, F. M. (2020). *Mudanças identificáveis no nível de Hemerobia resultantes da dinâmica do uso da terra na Bacia Hidrográfica do Rio Grande Comitán (Chiapas, México).* uma análise do período 1999–2019

Albuquerque, P. C. D., & Lima, F. A. X. (2023). Políticas públicas de sustentabilidade em Fortaleza, CE. *Interações (Campo Grande), 24*, 521–535.

Andrade, A. S. D. (2020). *Informação para políticas públicas de desenvolvimento em Rondônia* (Doctoral dissertation, Universidade de São Paulo).

Andrade, A. L. P. D. (2023). Bíblia e sustentabilidade: A consciência ecológica como paradigma de uma nova hermenêutica bíblica. *Perspectiva Teológica, 55*, 383–404.

Barbosa, R. A., & Queluz, M. L. P. (2021). *Ciência, tecnologia e história na poética antirracista de Rosana Paulino1.*

Batista, C. L. R., & Stoffel, J. (2022). Agroecologia e produção orgânica: características que distinguem e/ou aproximam os sistemas de produção sustentáveis. *COLÓQUIO-Revista do Desenvolvimento Regional, 19*(Edição Especial 1 (SOBER), março), 25–49.

Bampi, E., de Melo Santiago, G., da Luz, V. C., Garbin, E., Castamann, A., & Mossi, A. J. (2021). Soluções tecnológicas emergentes para uma agricultura sustentável: microrganismos eficientes. García, LMH Agroecologia: princípios e fundamentos ecológicos aplicados na busca de uma produção sustentável, Canoas: Mérida Publishers, 31–62.

Bortonio, N. D. S. (2023). *Associação de Bacillus sp. ao organomineral 06.30.00 na cultura do milho.*

Brandão, T. F. B. (2021). *Associação de mulheres" Resgatando sua História": organização social e gestão no rural nordestino¿ Porto da Folha¿ Sergipe.*

Brito, A. C. L. (2023). *O direito fundamental à alimentação adequada dos povos indígenas no Brasil e o cultivo de alimentos transgênicos em seus territórios.*

Coêlho, J. D. (2023). *Agricultura: Milho Produção e Mercado.*

Deslauriers, J. P., & Kérisit, M. (2023). *O delineamento de pesquisa qualitativa.*

Destefani, W. C. (2019). *Uma pesquisa etnomatemática com familiares e alunos do primeiro ano do ensino fundamental de uma escola agroecológica no município de Águia Branca–ES.*

Douradinho, R. S. (2023). *Aplicação de feixe de elétrons para esterilização de mosto obtido a partir da hidrólise de milho em caldo de cana energia, visando à produção de etanol* (Doctoral dissertation, Universidade de São Paulo).

Drehmer, A. C. B. (2020). *Sistemas agrários familiares de base agroecológica-1° Distrito de Canguçu/RS: uma análise multidimensional para o desenvolvimento rural* (Master's thesis, Universidade Federal de Pelotas).

Ferreira, C. O. (2021). *A Questão agroambiental no eixo temático TVT: estudo de caso na Escola Estadual Nova Itamarati.* Assentamento Itamarati II–Ponta Porã, MS.

García, L. M. H. (2021). Ecological principles and foundations applied in the search for sustainable production

Gonçalves, C. W. P. (2004). Geografia da riqueza, fome e meio ambiente: pequena contribuição crítica ao atual modelo agrário/agrícola de uso dos recursos naturais. *INTERthesis: Revista Internacional Interdisciplinar, 1*(1), 1.

Lima, R. S. D. (2019a). *Percepção das águas e do lugar na sub-bacia do rio das Lajes/Sergipe.*

Lima, T. S. (2019b). *Potencial do óleo essencial de cravo da índia sobre fungos fitopatogênicos em sementes de milho.*

Lima, P. C. D. (2023). *Novo rural na região imediata de Patos-PB: a construção de um livro paradidático para o 7° ano do ensino fundamental de geografia* (Master's thesis, Universidade Federal do Rio Grande do Norte).

Lopes, J. S. (2019). *Petróleo verde na Bahia.*

Lopes, E. V., & Padilha, N. S. (2019). Retrocessos no sistema de comunicação de riscos na rotulagem de agrotóxicos: a classificação da Anvisa. *Revista de Direito Ambiental e Socioambientalismo, 5*(2), 55–76.

Markovski, A. M. (2023). *A evolução da agroecologia no Município de Itatiba do Sul/RS: atores e ações.*
Medeiros, E. P. (2023). *Wetlands Construídos: economia, ecologia e sustentabilidade.*
Melo, R. L. D. M. (2022). *Sistemas agroflorestais e o legado de João Ambrósio de Araújo Filho: Subsídios para a criação do instituto de pecuária apropriada no município de Gurjão-Paraíba.*
Mendonça, M. V. (2019). *Corredor ecológico entre as áreas protegidas de Carajás e da Terra do Meio, Pará.*
Mezalira, S. M. (2023). *Da situação-limite agrotóxicos ao inédito-viável expresso pela agroecologia: compreensões e práticas de professores e estudantes de escolas do campo.*
Morais, P. D. P. (2019). *Fragmentação da paisagem e corredores ecológicos na região do MATOPIBA.*
Neto, B. S. (2020). As relações entre valor agregado e riqueza na Análise-diagnóstico de sistemas agrários. *Extensão Rural, 27*(4), 48–66.
Pieniz, F. G. (2021). *Valor agregado e desenvolvimento sustentável: o caso das agroindústrias rurais no município de Cerro Largo.*
Quevedo, M. G. (2020). *Estimativa dos ganhos econômicos com a rotação do arroz com a soja em Terras Baixas* (Master's thesis, Universidade Federal de Pelotas).
Reginatto, M. (2023). *Sistema plantio direto orgânico: rotações de cultura e controle mecânico de plantas espontâneas no sudoeste do Paraná.*
Reinach, M. H. D. M. (2020). *Controvérsias sobre a Dimensão Social da Agroecologia: Uma comparação entre os discursos políticos da FAO e da Via Campesina.*
Rocha, H. M. B. C. C. L. (2016). *Serviço Social e Ambiente: a sustentabilidade ecológica das comunidades socialmente vulneráveis.*
Santos, J. F. S. (2023). *Seleção de genótipos de milho e respostas fisiológicas ao déficit hídrico em região semiárida.*
Santos, L. S. C., & Garcia, R. P. M. (2020). Agroecologia: estratégia de luta para fortalecimento e resistência da Juventude Camponesa: estrategia para luchar por el fortalecimiento y la resistencia de la juventud campesina. *Kiri-Kerê-Pesquisa em Ensino, 1*(4).
Silva, J. R. A. D. (2019). *Milho não transgênico-Caracterização e formulação de produtos.*
Silva, D. I. D. C. (2020). *Significados atribuídos por agricultores familiares à prática da policultura na sua relação com a saúde e o ambiente.*
Silva, R. D. O. (2023). *Análise Socioeconômica e Agroambiental da Agricultura Familiar na comunidade rural de Fazendinha, Magalhães Barata, Pará.*
Silva, J. R. G. D., & Vergara, S. C. (2015). Análise comparativa acerca da utilização de desenhos na pesquisa sobre a criação do sentido da mudança organizacional. *Organizações & Sociedade, 9.*
Sorrini, T. B. (2022). *Restaurando o campus: envolvendo alunos e plantando árvores para a transformação ambiental da ESALQ/USP* (Doctoral dissertation, Universidade de São Paulo).
Torres, C. J. F. (2020). *Bases metodológicas para a inserção do conceito nexus água–energia–agricultura em modelos intersetoriais de planejamento e gestão.*
USDA—United States Department of Agriculture. (2023). *Production, supply and distribution (PSD) on line.*
Vallejo, L. G. (2022). *Sistemas agroecológicos: regime de informação de um campus universitário como um Living lab.*
Veiga, J. E. D. (2023). Uma linha estratégica de desenvolvimento agrícola. *Brazilian Journal of Political Economy, 12*, 249–266.
Vergara, S. C., & Peci, A. (2003). Escolhas metodológicas em estudos organizacionais. *Organizações & Sociedade, 10*, 13–26.
Vernini, A. A. (2019). *Energia incorporada na produção de biodiesel a partir da cultura do crambe.*
Vilanova, N. R. (2019). *Avaliação do crescimento e grau de herbivoria em plantas de milho cultivadas com bordaduras de feijão guandu, crotalaria e capim napier.*
Yin, R. K. (2005). *Estudo de caso: planejamento e métodos* (3rd ed.). Bookman.

Value Addition in Millets for Sustainable Food Security

2

Noopur Gautam and Chaudhary Mohd Kashim

Abstract

Nutri-cereals, often known as millet, are small grains used for food, feed, and fodder. Millets are a major food source in arid and semi-arid parts of the world. Millets are the precursors of nourishment essential for human wellness. In addition to a variety of vital nutritional components such as vitamins, minerals, and micronutrients, millet grains also include a significant quantity of anti-nutritional components. It is imperative to eliminate anti-nutritional components from the specific variety of millet before consuming it, as failure to do so poses significant health risks. Since ancient times, millet grains have been processed at the home level using a variety of methods to prepare them for human use. Soaking, heating, roasting, fermenting, cooking etc. are some of these processes. The goal of the intervention was to stop the millet population decline, which was mostly caused by difficulties preparing the grain as a direct food source. In order to minimize the annoyances and to create, improve, and standardize millet product technologies, efforts were made to diversify the millet product through processing technologies.

Keywords

Millet processing · Nutritional enhancement · Functional food · Food security

N. Gautam · C. M. Kashim (✉)
Institute of Food Technology, Bundelkhand University, Jhansi, Uttar Pradesh, India

C. B. Singh et al. (eds.), *Global Millets Production for a Sustainable Future*,
https://doi.org/10.1007/978-3-031-88157-2_2

Key Highlights

- Introduction: Millets and Their Nutritional Value.
- The study emphasizes on detailed review of existing literature on processing technologies for nutritional enhancement to improve health, value addition in millets for sustainable food production.
- Nutritional improvement via fortification or incorporating.
- Importance of value addition in increasing millet utilization and marketability.

2.1 Introduction

Millets are a forerunner of the nourishment needed for human wellness. Millets are essential crops for human nutrition and play a significant role in the global food system. Millets, often known as small grains, are farmed for their small kernels from grassy plants in the Poaceae family. Millet is a generic term used for small sized grains that form heterogeneous group and referred along with maize and sorghum as 'coarse cereals. There are many varities of millets. The four major types are pearl millet (*Pennisetum glaucum*) which compromises 40% of the world production, foxtail millet (*Setaria italica*), proso millet or white millet (*Panicum miliaceum*), and finger millet (*Eleusine coracana*). Minor millets include barnyard millet (Echinochloa spp.), kodo millet (*Paspalum scrobiculatum*), little millet (*Panicum sumatrense*), browntop millet (*Urochloa ramosa* = *Brachiaria ramosa* = *Panicum ramosum*), sorghum (*Sorghum spp.*) (ICRISAT 2007; FAO, 2012; Adekunle, 2012) (Fig. 2.1).

Millets are a valuable income crop for small farmers and may thrive in marginal soil conditions, benefiting almost one-third of the global population. Millets serve a vital function in food security as they can withstand harsh agro-climatic conditions Gautam et al. (2023). Millets have gained popularity in recent years due to their

Fig. 2.1 Shows different variety of millet grains along with their panicles

Table 2.1 Nutritional composition of millet

Nutrient	Percentage
Carbohydrate	60–70%
Protein	6–10%
Fat	1.5–5%
Fibre	12–20%
Minerals	2–4%

Source: V.T. Asharani et al. 2010

remarkable nutritional profile and ease of consumption in any form. Millets' nutritional and functional properties make them ideal for creating value-added products. Millets are mostly consumed by traditional customers due to the limited availability of ready-to-eat food products, despite their high nutritional value. Table 2.1 highlights the nutritional composition of millets.

2.1.1 Nutritional Composition of Millets

See Table 2.1.

2.2 Processing Technologies in Millets

Since millet crops are mostly processed and used on a domestic scale, their use is primarily limited to catmint production, depriving a big portion of the population of access to its vital nutrients. Different types of millets are consumed in India in the form of many goods such as porridge, chapati, laddu, machula, dubkee, churkani, and varied meals that are peculiar to several regions of the world. The anti-nutritional component found in millet grains is another typical cause of their low use. However, due to a lack of appropriate processing knowledge, there is limited consumption of these products, particularly in unconventional areas, at the necessary rate. To minimize or eliminate these anti-nutritional elements found in various millets and transform them into palatable form, a variety of traditional methods have been used. Before being consumed in households, several processes like soaking, germination, roasting, sprouting, fermentation, dehulling, and cooking have been done since ancient times Antony & Chandra (1998) (Fig. 2.2).

2.2.1 Soaking

Soaking is a typical treatment for grains that improves mineral bioavailability and reduces antinutritional substances such as phytic acid. Soaking affects phytate degradation, phytase activity, and mineral concentrations (e.g. iron, zinc) (Kate & Singh, 2021). (Fig. 2.3).

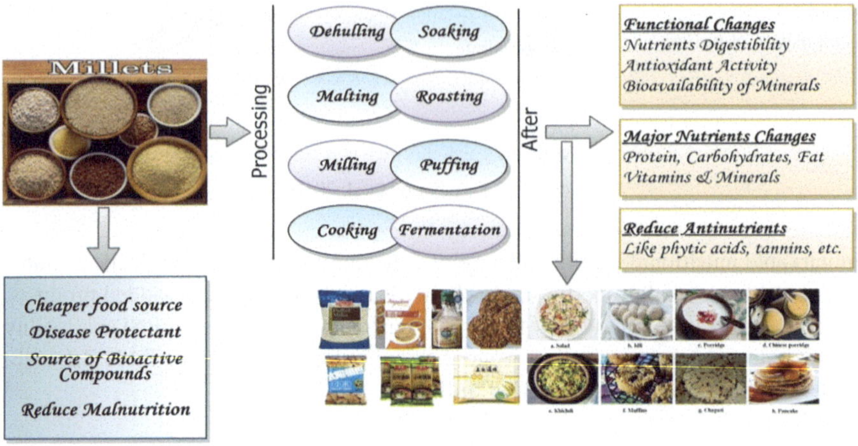

Fig. 2.2 Illustrates about different processing technologies that can be implied to reduce Anti-nutritional factors and produce value added products from these millets (Yousaf et al., 2021)

Fig. 2.3 Soaking of finger millet (ragi) grains in water

2.2.2 Dehulling/Dehusking/Decortication/Pearling

Traditionally, the process of pounding millet grains—dry, moistened, or wet—in a stone or wooden mortar with a pestle is used to dehusk them. A crucial component of the intended result is that the grains must have the right moisture content before they are dehulled. Therefore, proper detachment of fibrous bran and separation of germ and endosperm require about 10% moisture. Millets' non-edible parts were removed during the dehulling process, which also increased the nutrients' bioavailability, decreased the amount of anti-nutritional ingredients, and increased customer acceptance. or different kinds of millets, there are different kinds of dehullers, decorticators, and pearlers available. They are CIAE-Millet mills, roller mills, and abrasive type mills (Kate & Singh, 2021).

2.2.3 Cooking

The most important step that must be taken before any kind of millet grain is consumed by humans is cooking. At the molecular level, intricate modifications in physicochemical and functional characteristics, such as protein denaturation, starch gelatinization, released bound phenolic, and antioxidants, have been occurring. Simultaneously, a significant decrease in anti-nutritional components is also seen. For millets to retain their nutritional components particularly the minerals and phenolic contents—cooking time and temperature are critical factors.

2.2.4 Roasting

Roasting is a time-honoured method in which millet grains with the right level of moisture content are exposed to a conductive heat medium at a high temperature of 160–200 °C for varied amounts of time. Grain is kept at the proper moisture content by pre-soaking and then drying in the shade. During the roasting operation, the cotyledons of grain exposed to heat become shrink more than the outer husk and develop a gap in-between which aids the loosening of husk from the cotyledon. This loosening of husk further promotes husk removal during dehusking or pearling. Additionally, this heat treatment successfully neutralizes harmful and anti-nutritional ingredients found in millets while enhancing their distinct flavor (Kate & Singh, 2021).

2.2.5 Sprouting

Grain germination under regulated conditions is called sprouting. The latent embryo found in millet grains is triggered during the sprouting process by a specific moisture content and environmental factors (temperature and humidity). This germination period for millets ranges from 48 to 72 h at a room temperature of 25 to 30 °C. While sprouting greatly reduces the availability of anti-nutritional components like tannin and phytic acid, it increases the availability of minerals and vitamins found in millet. When sprouting for longer than 48 h, the dry weight of the source grain decreased but the nutrition did not significantly improve (Khamgaonkar 2012; (Mbithi-Mwikya et al. 2000)) (Fig. 2.4).

2.2.6 Fermentation

Throughout history, fermentation has been a widely employed preservation method for a wide range of food goods. Throughout the world, millets are used to produce a wide range of traditional meals. The physiochemical characteristics of the millet grains and the finished fermented product are considerably changed during the fermentation process. The reduction of antinutrients, increase in protein availability,

Fig. 2.4 Indicate the germination and sprouting process of pearl millet (bajra) and finger millet (ragi) Bhuvaneshwari et al., (2020)

increase in protein digestibility, and overall improvement in nutrition profile are the main benefits of the fermentation process (Kate & Singh, 2021).

2.2.7 Puffing

This widely used processing method is mostly used to create extended snacks that are ready to eat and other items made from any kind of grain. The millets being pre-soaked at the appropriate moisture content and subjected to hot sand in a 1:6 ratio at a high temperature (230–250 °C) for a brief period of time (20–30 s) in order to facilitate the popping/expansion process (Kate & Singh, 2021) (Fig. 2.5).

2.3 Effect of Different Processing Techniques on Different Bioactive Compounds of Millets

While various processing methods enhance the nutritional value of millet, they can occasionally have an adverse effect on its bioactive components. While optimizing the process parameters, consideration must always be given to both positive and negative effects on bioactive active substances. The overall amount of antioxidants in the millet grains and their activity had a major impact on the processing techniques such as soaking, decortication, malting, and cooking. The malting process, which also involves soaking, sprouting, and drying, includes germination, which can lower anti-nutrients like phytates, tannins, and oxalic acid and raise total phenolic content and nutritional value (Fig. 2.6).

Fig. 2.5 Showcases the puffed sorghum grains (https://www.eatthismuch.com/food/nutrition/puffed-millet,5784/)

Fig. 2.6 Millet grains source of bioactive compounds

Millets are said to undergo fermentation, which alters the ratio of nutritional to anti-nutritive components and increases the amount of biologically active components. Roasting, puffing, and parboiling are examples of heat treatment. While these treatments enhance the nutritional qualities of millets, such as total phenolic content, their impact on these qualities differs based on the type of heating technique employed. Cereal processing often involves dehulling and soaking. Millets are typically dehulled before further processing, and soaking can be used to minimize the presence of several antinutritional components in millets Wang et al. (2022).

2.4 Millet-Based Value-Added Food Products

2.4.1 Composite Flour

One of the most common traditional forms of millet eating for humans is as flour. It can be eaten as a single variety of millet flour or combined in the right amounts with other popular varieties of flour (Fig. 2.7).

The purpose of blending multiple flours is to obtain the appropriate functional, nutritional, and physicochemical properties in the flour. Consequently, in the case of wheat flour, this substitution is feasible at a level of 10–15%, similar to that of finger millet, proso millet, and barnyard millet, which have respective percentages of 10%, 15%, and 20%. As the amount of millet flour in the blend grows, the blend's ash content, gluten value, dough loaf volume, percentage of damaged starch, and protein all decrease. Roti, spaghetti, macaroni, vermicelli and noodles, pasta etc. can be prepared using composite flour (Kate & Singh, 2021; Khamgaonkar (2012)).

2.4.2 Bakery Product

Cereal flours are the primary ingredient in baked goods, which are then combined with additional ingredients. Different kinds of millets have been utilized to create well-known bakery goods like cakes, cookies, and biscuits. Because they are high in dietary fiber and low in gluten, these kinds of millet-based baking and confectionary

Fig. 2.7 Millet flour bulk of fibre and nutrition

goods are becoming more and more well-liked. People these days rarely have the time to prepare breakfast; instead, they usually eat cookies, bread, buns, cakes, and other baked goods Kumar et al. (2021). These days, consumers are demonstrating a preference for millet-based bakery items since they prefer healthier products over traditional dishes. In order to standardize cookies bread/buns and cakes, varying percentages of millet can be added, ranging from 10% to 50%. Refined wheat flour can be substituted with millet flour.

2.4.3 Cookies

Cookies are popular ready-to-eat product consumed by different age groups in a family. A rotating oven, automatic cookie maker, and planetary mixer can all be used to make millets cookies. Cookies are made using a blend of quality pearl millet, finger millet, and foxtail millet, sorghum flour, sugar, milk solids, trans fat, salt, and naturally occurring flavouring agents (Fig. 2.8).

2.4.4 Bread

Bread is made by combining flour, water, fat, salt, and yeast until the combination becomes dough, which is then baked into a loaf.

Bread is a ready-to-eat food. To make millet loaves, substitute kodo millet, proso millet, or foxtail millet flour in different amounts for the wheat in bread, and then

Fig. 2.8 Various types of millet cookies

Fig. 2.9 Millet bread

add premium yeast, trans-fat-free oil, salt, and sugar. To make bread, the dough is proofed and then baked. To make buns, dough is formed into round balls and baked (Fig. 2.9).

2.4.5 Flaked and Popped products

Popping of the grains is one of the age-old traditional practice for value addition. It is generally used for the preparation of snacks or breakfast cereals which are either plain or incorporated with spices, salts, and sweeteners Saraswathi & Hameed (2022). Cereal popped products and flakes are popular breakfast foods and at present they are mostly made from corn. By suitable processing it might be feasible to produce popped foods and flakes from millets. These Ready-To-Eat products are very popular, being crisp and friable in texture. The relatively smaller size and quick hydration of millets make them most suitable for the production of flakes and popped products (Fig. 2.10).

2.4.6 Traditional food

Idli, Dosa, Idiappam, Rotti, Pittu, Upma, Adai, porridge, Khakra, Paniyaram, and Chappathi are some examples of breakfast foods. Sweets include Kheer, Adhirasam, Kesari, Sweat Kolukattai, and Nutritious Balls. Vadai, Pakoda, Ribbon Pakoda, Omapodi, Murukku, Thattu Vadai, Hot Kolukattai, and Vadagam are among the snacks that can be made from millets (Fig. 2.11).

Fig. 2.10 Popped and flaked product made from millets

Fig. 2.11 Traditional products made from millets

2.4.7 Fermented Products

In India, many fermented millet products have been utilized from ancient times. These items have varied recipes depending on the socioeconomic status and geographic area. In addition to having a distinct flavor, fermented foods are higher in a variety of important macro- and micronutrients and have fewer antinutrients than raw grains or grains prepared using another method. Some of them are as follows (Fig. 2.12):

2.4.7.1 Dosa

Many residents of the area regularly consume dosa, a traditional South Indian dish. It is a fermented pancake made from rice and lentil batter. Dosa can, however, be prepared in a variety of ways using a variety of grains, including millet. Those seeking a gluten-free and healthier alternative often turn to dosa made from fermented millet batter.

2.4.7.2 Rabadi

Native to India, the lactic fermented milk beverage known as "rabadi" is a favorite in the country's northwestern semiarid regions. Indian dairy dessert known as millet-based fermented rabadi is made with millet, milk, and jaggery. In India,

Fig. 2.12 Dhokla prepared from foxtail millet (https://cookwithrenu.com/foxtail-millet-dhokla-steamed-millet-cake/)

millet has been grown for many centuries and is a healthy, gluten-free grain. Milk is reduced to about one-third of its original volume by simmering it until it thickens, creating the sweet, and creamy dessert known as rabadi Patil et al. (2023).

2.4.7.3 Idli

In various regions of India, pearl millet, sometimes referred to as bajra, is a widely consumed cereal grain. Making idlis, a common breakfast food in South India, is one of the most popular methods to consume pearl millet. Pearl millet based idlis, in contrast to conventional rice-based idlis, are becoming more and more well-liked because of their numerous health advantages. Idlis prepared from pearl millet are a fermented food created from the ingredients urad dal (split black gramme) and fenugreek seeds Patil et al. (2023) (Fig. 2.13).

2.4.7.4 Koko

A key ingredient in West African cuisine, Koko is a tasty and wholesome fermented food. It is a well-liked option for breakfast, snacks, and other meals thanks to its distinctive flavor and adaptability Patil et al. (2023).

2.4.7.5 Jandh

Traditional fermented foods are typically unique to individual cultures and geographical regions. Jandh, a mildly sweet and acidic style of beer beverage is one of Nepal's main traditional alcoholic products. Jandh is a finger millet (koko or marua) fermentation product that is occasionally combined with a little amount of wheat or maize Patil et al. (2023).

Fig. 2.13 Healthy traditional fermented products from millets

2.5 Extruded Products

The cooking method known as extrusion involves heating grains to a high temperature and quickly grinding them before feeding them into an apparatus known as an extruder. The apparatus maintains the proper temperature, pressure, residence time, and feed rate settings. It is possible to create a wide range of distinct kinds and attributes of millet-based extruded goods using variable grain compositions, feed rates, cooking temperatures, pressures, and residence times.

References

Adekunle, A. A. (2012). *Agricultural innovation in sub-saharan africa: experiences from multiple-stakeholder approaches*. Forum for Agricultural Research in Africa. ISBN 978-9988-8373-2-4.

Antony, U., & Chandra, T. S. (1998). Antinutrient reduction and enhancement in protein, starch, and mineral availability in fermented flour of finger millet (Eleusine coracana). *Journal of Agricultural and Food Chemistry, 46*(7), 2578–2582.

Asharani, V. T., Jayadeep, A., & Malleshi, N. G. (2010). Natural antioxidants in edible flours of selected small millets. *International Journal of Food Properties, 13*(1), 41–50. https://doi.org/10.1080/10942910802163105

Bhuvaneshwari, G., Nirmalakumari, A., & Kalaiselvi, S. (2020). Impact of soaking, sprouting on antioxidant and anti-nutritional factors in millet grains. *Journal of Phytology, 12*, 62–66. https://doi.org/10.25081/jp.2020.v12.6384

Cook With Renu. (n.d.). Retrieved from https://cookwithrenu.com/foxtail-millet-dhokla-steamed-millet-cake/

Eat This Much (n.d.). Retrieved from https://www.eatthismuch.com/food/nutrition/puffed-millet,5784/

FAO (Food and Agriculture Organization). (2012). Economic and Social Department: The Statistical Division. Statistics Division 2012. Retrieved September 29, 2012, from http://faostat.fao.org/site/567/DesktopDefault.Aspx?PageID=567

Gautam, N., Singh, M., Arya, P., Kumar, R., Bhargava, P., & Gupta, A. (2023). *Goodness of millets in daily meals through easy & healthy recipes* (1st ed.). Bundelkhand University. ISBN 978-81-958954-8-9.

ICRISAT. (2007). *2007 Annual report*. International Crops Research Institute for the Semi-Arid Tropics.

Kate, A., & Singh, A. (2021). Processing technology for value addition in Millets. *In Springer eBooks* (pp. 239–254). https://doi.org/10.1007/978-981-16-0676-2_11

Khamgaonkar, S. G. (2012). *Optimization of processing parameters for development of underutilized cereals and legumes based composite flour* (Unpublished Thesis, Govind Ballabh Pant University of Agriculture and Tech).

Kumar, A., Tripathi, M. K., Joshi, D., & Kumar, V. (2021). *Millets and millet technology*. Springer Nature. ISBN: 978-981-16-0675-5.

Mbithi-Mwikya, S., VanCamp, J., Yiru, Y., Huyghebaert, A. (2000) *Nutrient andantinutrient changes in f inger millet (Eleusine coracan) during sprouting*. LWT-Food Sci Technol 33(1):9–14.

Patil, A. G., et al. (2023). Millets based fermented products: A review. *Acta Scientific Nutritional Health, 7*(6), 69–81.

Saraswathi, R., & Hameed, R. S. (2022). Value addition influenced in millet product—A review. *Journal of Nutrition and Food Science, 10*(1), 1161.

Wang, H., Fu, Y., Zhao, Q., Hou, D., Yang, X., Bai, S., Diao, X., Xue, Y., & Shen, Q. (2022). Effect of different processing methods on the millet polyphenols and their anti-diabetic potential. *Frontiers in Nutrition, 9*, 780499. https://doi.org/10.3389/fnut.2022.780499

Yousaf, L., Hou, D., Liaqat, H., & Shen, Q. (2021). Millet: A review of its nutritional and functional changes during processing. *Food Research International, 142*, 110197. https://doi.org/10.1016/j.foodres.2021.110197

The Role of Millets in Maharashtra, India: Examining Production, Consumption Patterns, and Their Significance in Influencing Food Security

3

Anil S. Maheshwari and Vijay Kulkarni

Abstract

This paper explores the several functions of millets in India, providing a thorough analysis of their production and consumption trends as well as the resulting effects on food security. In addition, a thorough examination of consumption trends provides insight into the nutritional choices and dietary habits of the Indian population. This study delves into the importance of millets in relation to food security and highlights its capacity to mitigate nutritional inadequacies, bolster climate variability tolerance, and promote sustainable farming methods. The results are intended to educate stakeholders and policymakers on the critical role that millets play in supporting diverse diets, strengthening India's food security framework, and advancing agricultural sustainability. Millets are resilient crops that are also a rich source of nutrients. Since the government placed more of an emphasis on fine grains and did not have appropriate policies for millets, the cultivation of millet has decreased throughout the nation. The area, productivity, and production of ragi, bajra, jowar, and small millets were the main subjects of the study. Pune, Mumbai, Nagpur, and Nashik are the states that produce the tiniest millets, jowar, bajra, and ragi, correspondingly. The area planted to small millets, jowar, bajra, and ragi decreased between 2018 and 2022 at a compound annual growth rate (CAGR) of 3.60%, 1.86%, 1.27%, and −0.60%, respectively. The output of the small millets and jowar has fallen with growth rates of −2.89% and −0.63%, respectively from 2019–20 to 2021–22 and ragi production increased with growth rate of 1.68%. According to decomposition analysis, the yield effect boosted ragi and bajra production, while the areas effect enhanced

A. S. Maheshwari (✉)
School of Engineering, Ajeenkya DY Patil University, Pune, Maharashtra, India

V. Kulkarni
Director Corporate & Executive Program, Ajeenkya DY Patil University, Pune, Maharashtra, India

C. B. Singh et al. (eds.), *Global Millets Production for a Sustainable Future*,
https://doi.org/10.1007/978-3-031-88157-2_3

jowar and small millets output. Over time, less people have been eating these millets in both rural and urban areas.

Keywords

Millets · Maharashtra · Production · Food security · Urban areas · Rural

3.1 Introduction

Millets have become important players in the complex fabric of India's agricultural environment, significantly influencing the country's perspective on food security. For decades, India's agriculture has relied heavily on millets, a family of small-seeded grasses that feed millions of people throughout the subcontinent. This introduction explores the many facets of millets in India, revealing the complex interactions between consumption, production, and the overall role these grains play in bolstering food security in the nation. Due to its capacity to adapt to a wide range of agroclimatic conditions, millets, which include a variety of grains like sorghum, finger millet, pearl millet, and foxtail millet, have been cultivated for a long time in India. With their minimal water requirements and ability to withstand severe environments, millets are an excellent option for growing in areas with unpredictable rainfall patterns and limited resources. As a result, millet farming has permeated several states' agricultural methods and significantly increased the amount of food produced in the country as a whole.

An analysis of millet production in India reveals a changing terrain characterized by cropping patterns and regional variances. The agro-ecological diversity present in the nation is reflected in the specialization of several states in the production of particular millet varieties. Comprehending these subtleties is crucial to grasping the complexities of millet production and its consequences for the larger agricultural industry. Beyond the fields of production, a closer look at millets' consumption habits reveals a rich culinary history closely linked to these grains. Once deemed traditional and restricted to particular populations, millet is making a comeback in contemporary diets. Millets are making their way into mainstream diets and bridging ethnic divides thanks to growing knowledge of their nutritional advantages and wide range of culinary uses. This change in consumption habits has significant effects on food security since adding millets to a variety of diets increases dietary diversity and nutritional resilience.

It is impossible to overestimate the importance of millets in affecting food security. The ability of millets to withstand harsh temperature conditions makes them an invaluable tool for maintaining a steady supply of food. Furthermore, millets are viable options in the face of changing agricultural landscapes because of their capacity to flourish in marginal soils and little input requirements. The goal of this investigation of the function of millets in India is to make sense of the complex web of their production, consumption, and overall importance in relation to food security. We hope to shed light on the little-known history of millets and their revolutionary potential to reshape India's agricultural and food supply in the next years by exploring these features.

3.1.1 Historical Significance of Millets in India

The cultivation of these hardy grains was the basis for the prosperity of ancient civilizations, which is why millets have such historical significance in India. Millets, including finger millet, foxtail millet, pearl millet, and sorghum, have been grown for thousands of years in the Indian subcontinent and are a staple food for many different kinds of people living there. Ancient books and traditions in Indian history make reference to the cultivation of millets. Millets have a long history in Indian agriculture; archaeological evidence indicates that they were one of the main crops grown during the Indus Valley Civilization. The adaptability of millets to diverse ecological niches made it possible for them to be grown throughout the subcontinent.

During the mediaeval period, millets continued to play a key part in Indian agriculture, with their cultivation spreading across many kingdoms and empires. Because millets are hardy, they may be grown in arid and semi-arid locations, which greatly improves the food security of the communities that live there. Millets' ability to withstand harsh weather conditions has cemented their standing as a staple crop throughout unpredictable and limited times. In ancient Indian communities, millet had cultural and religious value as well. They were included into social customs, culinary customs, and religious acts in addition to being considered as food. Millets have a significant role in the nation's cultural fabric because they are frequently used in a variety of culinary preparations for festivals and festivities.

3.1.2 Shift in Agricultural Landscape

The agricultural landscape throughout the world and in certain regions—like India—has undergone a significant transformation in the past several years. The need for resilient and sustainable agricultural practices is becoming increasingly apparent, and cropping patterns are changing in tandem with technology developments to characterize this transition. The changing climate, shifting consumer tastes, and the need to provide food security for an expanding population are all contributing to the transformation of India's agricultural environment.

Crop diversity is one significant feature of this change. of the past, the agricultural landscape of India was dominated by a few primary crops, such as rice and wheat. But crop diversity has become more and more important as a reaction to the problems brought on by soil degradation, water shortages, and climate change. Oilseeds, pulses, and millets are just a few of the crops that farmers are experimenting with as alternatives to the conventional mainstays. By diversifying, one can lessen the hazards involved in monocropping and increase resistance to shifting environmental conditions. A major factor in the agricultural landscape's transformation has been the development of technology. Enhancing efficiency and production have been the results of integrating digital technologies, using genetically modified crops, and implementing precision farming. Farmers can use resources more wisely, use inputs more efficiently, and lessen the effects of climatic variability with the help of satellite images, data analytics, and precision agriculture tools.

3.2 Research Objectives

- To Examine the causes of the decline in millet production, paying particular attention to governmental regulations, consumer preferences, and environmental factors.
- To Examine regional variations in millet production, taking into account socio-economic, climatic, and agricultural factors, with an emphasis on Pune, Mumbai, Nagpur, and Nashik.
- To Examine the patterns of millet consumption in both urban and rural settings, taking into account variables including shifting dietary preferences, societal changes, and knowledge of the grain's nutritional advantages.
- To Examine the patterns of millet consumption in both urban and rural settings, taking into account variables including shifting dietary preferences, societal changes, and knowledge of the grain's nutritional advantages.

3.3 Review of Literature

A comprehensive study of millets, with a primary focus on breeding, nutrition, food security, and sustainability, was published in the Journal of Soil Science and Plant Nutrition in 2023 by Bandyopadhyay et al. The writers skillfully compile a broad spectrum of literature, offering scholars, decision-makers, and practitioners an invaluable tool. The emphasis on breeding strategies fits with the necessity to enhance millet types for increased production and durability. Moreover, the investigation of nutritional elements highlights millets' potential to alleviate malnutrition, a growing global concern. The writers also explore the vital relationship between millets and sustainability and food security, illuminating the diverse ways in which millets contribute. For individuals looking for a more nuanced understanding of the opportunities and difficulties related to millets in the twenty-first century, this study is a helpful resource.

The Asian Journal of Agricultural Extension, Economics & Sociology study by Chaudhary et al. (2023) examines the production, consumption, and effects of millets on food security with a particular focus on India. The writers offer a careful examination of millets' place in India's intricate network of nutritional and agricultural dynamics. The article provides significant insights for policymakers and stakeholders who are involved in forming the agricultural landscape of the nation by clarifying the patterns of production and consumption. The focus on millets' effects on food security highlights how important a role they play in maintaining a resilient and diverse food chain. The importance of millets in the Indian agricultural and nutritional framework is explored in this study at a relevant moment.

Datta's (2023) contribution to the State of India's Livelihoods places millets in a unique position within the larger context of livelihoods. The author gives the discussion of millet cultivation a socio-economic component by analyzing how millets support livelihoods. This paper is particularly useful since it examines the ways in which millets improve local economies. Datta's research offers crucial insights into the potential of millets as a means of promoting sustainable livelihoods as well as serving as a source of nutrition. Our knowledge of the many advantages of millet farming is deepened by the inclusion of millets in the conversation about livelihoods.

The work of Ghosh and Ramana (2023), which is discussed in a book chapter, adopts a novel strategy by looking at how millet consumption and sustainable practices interact in Southern Orissa. With an emphasis on the financial elements of millet consumption, the writers explore the consequences of encouraging sustainable decisions. Through the inclusion of millets in the broader framework of ethical consumer behavior, this chapter offers important new perspectives on how economic variables influence the selection of sustainable foods. Our comprehension of how millets might be incorporated into regional economies and consumer behaviors is deepened by the incorporation of a regional perspective. The multidisciplinary approach provides a nuanced perspective on millets' potential to support ecologically conscious consumption by fusing sustainability and marketing studies.

With an emphasis on finger millet in India, Grovermann et al.'s (2018) study, which was published in Agriculture, offers a thorough approach to evaluating the economic repercussions of underutilized crops. By analyzing productivity in the context of climatic resilience, food security, and nutrition, the authors give a complete examination of the economic viability of finger millet production. The study links economic factors with climatic resilience and nutritional results, addressing the wider implications of promoting underutilized crops for sustainable agriculture. Policymakers and agricultural stakeholders looking for evidence-based approaches to improve food security in the face of climate change may find this research to be very pertinent.

The determinants and policy implications of adding millets to India's food and nutrition security framework are the main topics of Islam and Manaloor's (2021) publication in the Journal of Nutrition and Food Security. The writers provide light on the socioeconomic, cultural, and policy aspects of the factors influencing millets' acceptance through their critical analysis. For academics and policymakers looking to create focused initiatives to encourage millet consumption for better nutrition, this study provides insightful information. The focus of the article on policy implications highlights the necessity of utilising millets as a crucial element of food security policies in India through a comprehensive and integrated approach.

3.4 Research Methodology

3.4.1 Study Design

The study uses both exploratory and descriptive-analytical methodologies in its dual approach. In order to find new advancements and trends in the selected topic, it strives for a comprehensive examination and deep grasp of the research subject. This method offers a thorough and multifaceted viewpoint on the study issue.

3.4.2 Data Collection Methods

- Primary Data
  - Online poll with standardized format using Google Forms.
  - 160 respondents participated, and 150 complete feedbacks were chosen for analysis.
  - Random participant selection to ensure a representative sample.
- Secondary Data
  - Derived from trustworthy sources like scholarly journals and government reports.
  - The dual-source approach improves the reliability and depth of the investigation.

3.4.3 Sampling Strategy

150 people are purposefully chosen at random from a pool of 160 survey respondents as part of the study's sampling approach. This deliberate randomization strengthens the external validity of the study and improves its key components, making the results more broadly applicable.

3.4.4 Data Analysis

Utilising quantitative analysis of primary data and combining secondary data for a thorough understanding, the data analysis takes a comprehensive approach. The objective is to obtain significant insights from primary and secondary sources in order to guarantee a comprehensive understanding of the study topic.

3.4.5 Survey Instrument

The Google Forms survey instrument has been meticulously created with an emphasis on clarity, relevance, and connection with the objectives of the research. Careful preparation guarantees effective data collection, which supports the methodological integrity of the study and the extraction of relevant, high-quality data for analysis.

3.4.6 Ethical Considerations

The study's ethical considerations give priority to voluntary involvement, privacy, and anonymity. Participants' identities are protected by stringent processes, and their rights and wellbeing are guaranteed by obtaining their voluntary, informed agreement. This emphasizes a careful adherence to ethical standards and scientific integrity throughout the investigation.

3.5 Data Analysis and Interpretation

The area, productivity, and production of tiny millets in Maharashtra are broken down by city in Fig. 3.1 (Table 3.1).

The information supplied shows the agricultural statistics for different Maharashtra cities, emphasizing the area under cultivation, yield, and productivity expressed in kg/Aurangabad makes up 13% of the whole area examined, with 20,000 ha under cultivation. The productivity is 10 kg/ha, or 6.67% of total

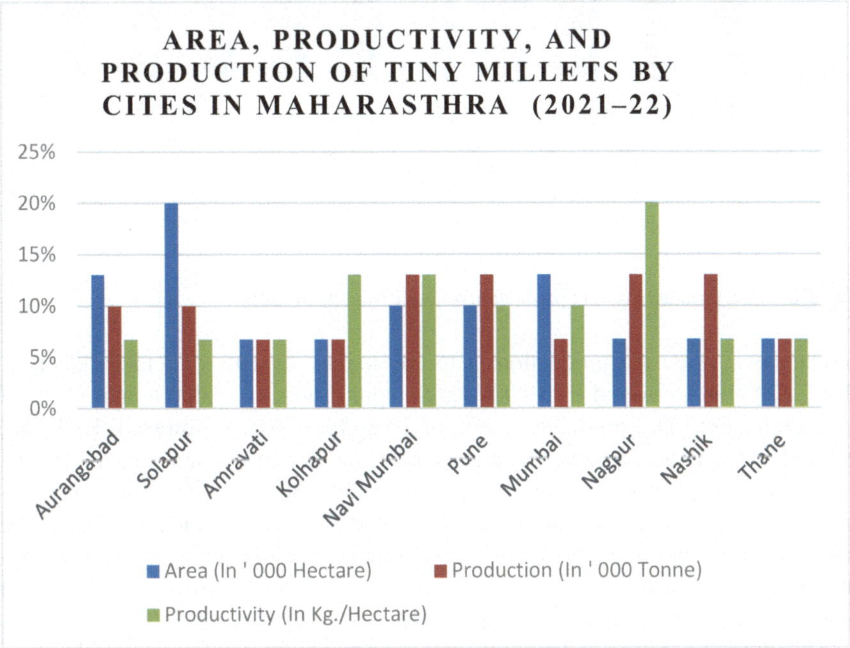

Fig. 3.1 Area, productivity, and production of tiny millets by state in India

Table 3.1 Area, productivity, and production of tiny millets by state in India

Cities	Area ('000 ha)	Production ('000 tonne)	Productivity (kg/ha)
Aurangabad	20 (13%)	15 (10%)	10 (6.67%)
Solapur	30 (20%)	15 (10%)	10 (6.67%)
Amravati	10 (6.67%)	10 (6.67%)	10 (6.67%)
Kolhapur	10 (6.67%)	10 (6.67%)	20 (13%)
Navi Mumbai	15 (10%)	20 (13%)	20 (13%)
Pune	15 (10%)	20 (13%)	15 (10%)
Mumbai	20 (13%)	10 (6.67%)	15 (10%)
Nagpur	10 (6.67%)	20 (13%)	30 (20%)
Nashik	10 (6.67%)	20 (13%)	10 (6.67%)
Thane	10 (6.67%)	10 (6.67%)	10 (6.67%)

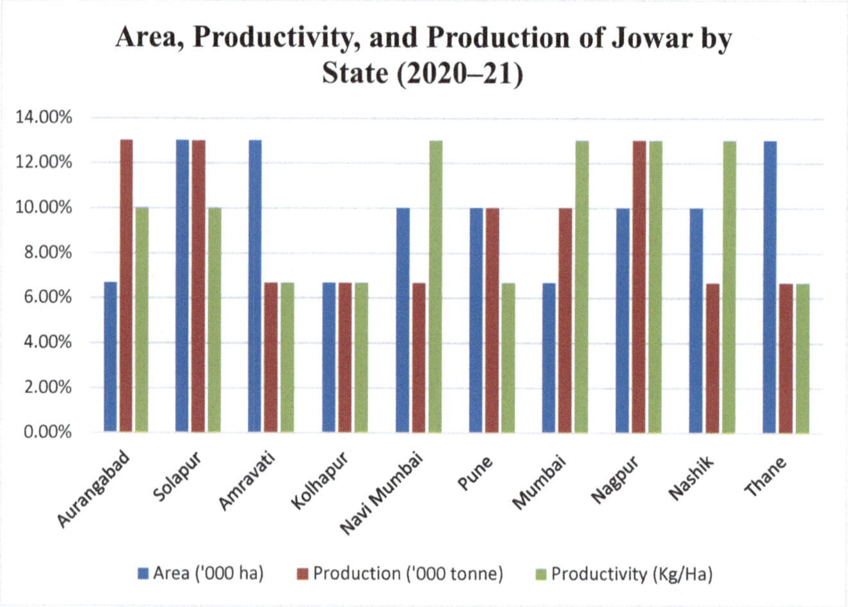

Fig. 3.2 Area, productivity, and production of jowar by state in India

productivity, while the production is 15,000 tonnes, or 10% of total production. With a 20% area covered by 30,000 ha, Solapur produces 15,000 tonnes (10% of total production) at a productivity rate of 10 kg/ha (6.67%). Amravati, Kolhapur, Navi Mumbai, Pune, Mumbai, Nagpur, Nashik, and Thane all contribute to the agricultural landscape in Maharashtra. Nagpur is notable for having the greatest productivity rate of 30 kg/ha, accounting for 13% of the overall production, 20% of the measured area, and a notable agricultural output. All things considered, these figures shed light on the variety of farming practices and the disparities in output among Maharashtra's cities (Fig. 3.2) (Table 3.2).

The information provided provides an overview of important agricultural indicators for different Maharashtra cities, including the area planted to crops, the amount

Table 3.2 Area, productivity, and production of jowar by state in India

Cities	Area ('000 ha)	Production ('000 tonne)	Productivity (kg/ha)
Aurangabad	10 (6.67%)	20 (13%)	15 (10%)
Solapur	20 (13%)	20 (13%)	15 (10%)
Amravati	20 (13%)	10 (6.67%)	10 (6.67%)
Kolhapur	10 (6.67%)	10 (6.67%)	10 (6.67%)
Navi Mumbai	15 (10%)	10 (6.67%)	20 (13%)
Pune	15 (10%)	15 (10%)	10 (6.67%)
Mumbai	10 (6.67%)	15 (10%)	20 (13%)
Nagpur	15 (10%)	20 (13%)	20 (13%)
Nashik	15 (10%)	10 (6.67%)	20 (13%)
Thane	20 (13%)	10 (6.67%)	10 (6.67%)

produced, and the productivity expressed in kg/ha. Aurangabad makes up 6.67% of the entire studied area, with 10,000 ha under cultivation. The productivity rate of 15 kg/ha is achieved by the city, indicating a noteworthy output, and it provides 13% of the total production, or 20,000 tonnes. With 20,000 ha (13% of the entire area), Solapur reports 20,000 tonnes of produce (13% of the total production) with a productivity of 15 kg/ha. Amravati, Kolhapur, Navi Mumbai, Pune, Mumbai, Nagpur, Nashik, and Thane each play a role in the agricultural landscape of Maharashtra. Among the examined areas, Nagpur occupies 20%, making a notable contribution of 13% to the total production and productivity at 20 kg/ha. These figures, which highlight differences in cultivation methods, product yields, and productivity levels among Maharashtra's cities, provide insights into the agricultural dynamics. The information is essential for comprehending and organizing local agriculture policies and resource distribution (Fig. 3.3) (Table 3.3).

The above data provides agricultural statistics, including the cultivated area, production volume, and productivity expressed in kilogrammes per hectare, for a number of Maharashtran cities. Both Aurangabad and Solapur have comparable agricultural profiles: they both produce 15,000 tonnes (10% of total production), span 20,000 ha (13% of the total area), and have productivity rates of 10 kg/ha (6.67%). Comparable data are also shown by Kolhapur and Amravati, where there are 10,000 ha of cultivated land (6.67%), 20,000 tonnes of production (13%), and 20 kg/ha of productivity (13%). Thane, Navi Mumbai, and Pune all have unique patterns. Navi Mumbai produces 10,000 tonnes (6.67%) at a productivity rate of 20 kg/ha (13%), utilising a 10,000-ha area (6.67%). Pune's production is 10,000 tonnes (6.67%) at a productivity rate of 20 kg/ha (13%), encompassing 20,000 ha (13%). Thane yields 20,000 tonnes (13%) and 10 kg/ha (6.67%) of productivity out of a 20,000 ha area (13%). Mumbai, Nagpur, and Nashik are three cities with diverse agricultural profiles. With a productivity rate of 15 kg/ha (10%), Mumbai covers 10,000 ha (6.67%) and produces 10,000 tonnes (6.67%). Nagpur produces 15,000 tonnes (10%) at a productivity rate of 15 kg/ha (10%) on an area of 15,000 ha (10%). With a productivity rate of 10 kg/ha (6.67%), Nashik, which covers 15,000 ha (10%), produces 15,000 tonnes (10%) (Fig. 3.4).

Table 3.4 showed how the nation's consumption of millet changed between 2018–20 to 2021–22.

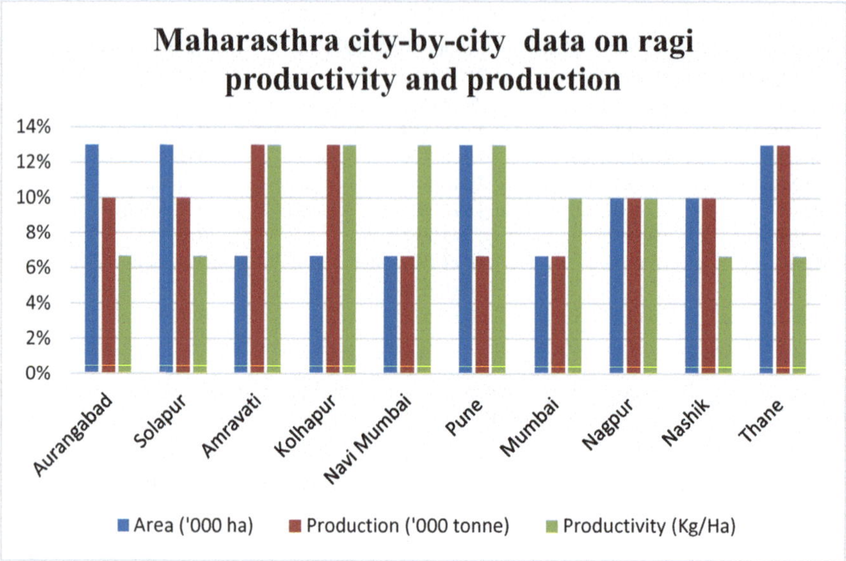

Fig. 3.3 Area, productivity, and production of bajra by state in India

Table 3.3 Area, productivity, and production of bajra by state in India

Cities	Area ('000 ha)	Production ('000 tonne)	Productivity (kg/ha)
Aurangabad	20 (13%)	15 (10%)	10 (6.67%)
Solapur	20 (13%)	15 (10%)	10 (6.67%)
Amravati	10 (6.67%)	20 (13%)	20 (13%)
Kolhapur	10 (6.67%)	20 (13%)	20 (13%)
Navi Mumbai	10 (6.67%)	10 (6.67%)	20 (13%)
Pune	20 (13%)	10 (6.67%)	20 (13%)
Mumbai	10 (6.67%)	10 (6.67%)	15 (10%)
Nagpur	15 (10%)	15 (10%)	15 (10%)
Nashik	15 (10%)	15 (10%)	10 (6.67%)
Thane	20 (13%)	20 (13%)	10 (6.67%)

The table shown delineates the cultivation trends of many millets, namely Ragi, Bajra, Jowar, and Small Millets, in both rural and urban areas from 2018–2019 to 2021–2022. In both rural and urban areas, the values indicate the proportion of total cropped area allotted to each type of millet. Jowar was the most common crop in both rural (0.41%) and urban (0.44%) areas in 2018–2019, while Bajra cultivation was somewhat greater in urban areas (0.39%) than in rural areas (0.40%). Ragi had a similar pattern, with higher cultivation in rural regions (0.44%) compared to urban areas (0.39%), and small millets were more commonly farmed in rural areas (0.41%) than in urban areas (0.30%). There were several noticeable shifts in the patterns of agriculture from 2019 to 2020. Jowar had a significant decline in urban areas (0.15%) but an increase in rural farming (0.51%). In terms of percentage terms, Bajra held steady in both rural (0.43%) and urban (0.41%) areas. On the other hand, small millets and ragi saw a fall in both urban and rural areas.

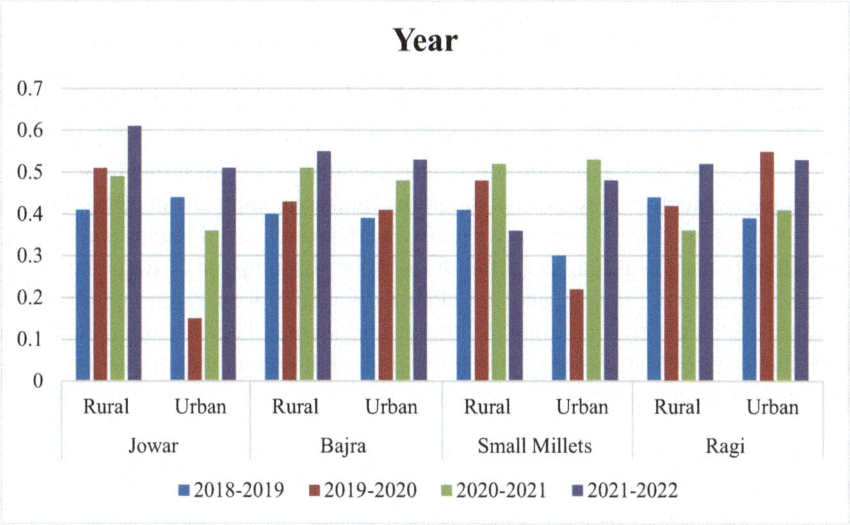

Fig. 3.4 Maharashtra per capita millet consumption (kg/capita/month)

Table 3.4 Maharashtra per capita millet consumption (kg/capita/month)

	Jowar		Bajra		Small millets		Ragi	
Year	Rural	Urban	Rural	Urban	Rural	Urban	Rural	Urban
2018–2019	0.41	0.44	0.40	0.39	0.41	0.30	0.44	0.39
2019–2020	0.51	0.15	0.43	0.41	0.48	0.22	0.42	0.55
2020–2021	0.49	0.36	0.51	0.48	0.52	0.53	0.36	0.41
2021–2022	0.61	0.51	0.55	0.53	0.36	0.48	0.52	0.53

There was a change in the cultivation environment in 2020–2021. Jowar's increasing tendency persisted in both urban (0.36%) and rural (0.49%) regions. Both urban (0.48%) and rural (0.51%) cultivation increased in Bajra. Ragi had a greater percentage in urban areas (0.41%) than in rural areas (0.36%), and small millets showed an increase in urban areas (0.53%) relative to rural areas (0.52%). In the most recent data for 2021–2022, Jowar continues to dominate in both rural (0.61%) and urban (0.51%) areas. Bajra cultivated in both rural (0.55%) and urban (0.53%) areas with a very balanced pattern. While Ragi exhibited stable cultivation percentages in both rural and urban areas, Small Millets showed a rise in both rural (0.52%) and urban (0.53%) locations.

3.6 Conclusion and Recommendation

In order to examine millet cultivation in Maharashtra cities—Aurangabad, Solapur, Amravati, Kolhapur, Navi Mumbai, Pune, Mumbai, Nagpur, Nashik, and Thane—the study used a dual methodology that included exploratory and descriptive-analytical approaches. The study's depth and dependability were increased by secondary data from reliable sources and primary data collected through an online Google Forms survey with 160 respondents (150 of whom were analyzed). The unique contribution of Nagpur, with its 30 kg/ha productivity rate, 20% area coverage, and 13% overall production, is one of the discoveries that should be noted. Certain millet investigations, like the ones for Jowar and Bajra, showed that different cities had different agricultural profiles, with differences in growing practices, yields, and productivity levels. Data on millet intake per capita from 2018–2019 to 2021–2022 showed shifting dietary patterns, with swings and dominance of Jowar, a balanced pattern of Bajra, and differences in the consumption of Ragi and Small Millets. Protection of privacy and voluntary participation were given top priority in the study's ethical considerations. All things considered, the study offers thorough understandings of Maharashtra's millet farming, highlighting the necessity of well-informed regional governance.

3.6.1 Recommendation

Recommendations based on the study include promoting best practices from Nagpur's distinctive agricultural model to enhance productivity in other regions, tailoring agricultural strategies to the specific needs of each city, implementing initiatives to increase awareness and consumption of millets, and maintaining a commitment to ethical considerations in future research and policymaking. These recommendations are based on the findings of the study.

References

Bandyopadhyay, T., Singh, R. K., Ramesh, P., Maurya, J., & Prasad, M. (2023). The promise of millets in the twenty-first century: Emphasis on breeding, nutrition, food security and sustainability. *Journal of Soil Science and Plant Nutrition, 23*(1), 628–637.

Chaudhary, J., Shelar, R., Thakur, K., & Singh, R. (2023). Millets in India: Production, consumption and impact on food security. *Asian Journal of Agricultural Extension, Economics & Sociology, 41*(8), 151–162.

Datta, S. (2023). *Livelihoods in millets. State of India's Livelihoods* (p. 95).

Ghosh, N., & Ramana, D. V. (2023). Millet and money promoting sustainable consumption in southern Orissa. In *Dealing with socially responsible consumers: Studies in marketing* (pp. 207–229). Springer Nature Singapore.

Grovermann, C., Umesh, K. B., Quiédeville, S., Kumar, B. G., & Moakes, S. (2018). The economic reality of underutilised crops for climate resilience, food security and nutrition: Assessing finger millet productivity in India. *Agriculture, 8*(9), 131.

Islam, S., & Manaloor, V. (2021). Millets for food and nutrition security in India: determinants and policy implications. *Journal of Nutrition and Food Security, 6*(2), 185–192.

Millet as a Nutri-Cereal for the Twenty-First Century: Celebrating the International Year of Millet 2023

4

Shikha Singh

Abstract

Due to the minerals, protein, and cell reinforcements that they contain, millets are sturdy cereals that can give a choice that is both financially savvy and nutrient-rich. Also, they can guarantee food security with their presence. As entire grains, millet arrives in a wide assortment of assortments, every one of which has an alternate kind of fiber and an alternate measure of fiber. This reality features the significance of rejuvenating millet crops. Then again, this fiber is a fundamental part in the guideline of gastrointestinal capability, glucose levels, and the digestion of lipids. Over a time of five years (2015–2020), this study explored the creation designs and nutritional organization of five fundamental grains in India: wheat, bajra, rice, maize, and grain. The review was led in India. There was a general expansion underway of all cereals that were inspected, aside from rice, with grain showing the most consistent expanding pattern. Specifically, foxtail millet arose as the victor as far as its protein content, while pearl millet was the champ regarding the two its fiber and iron substance.

Keywords

Millet · Nutri-cereal · Twenty-first century · International Year of Millet 2023 · Food and Agriculture Organization · Indian Council of Medical Research · United States Department of Agriculture

S. Singh (✉)
School of Management, Ajeenkya D Y Patil University, Pune, Maharashtra, India

C. B. Singh et al. (eds.), *Global Millets Production for a Sustainable Future*,
https://doi.org/10.1007/978-3-031-88157-2_4

4.1 Introduction

Millet is an aggregate term that alludes to various types of oat grass that have tiny grains. The significant millets, like sorghum and pearl millet, have been recognized from the more modest grain millets, which incorporate finger millet (ragi), foxtail millet (kangni), kodo millet (kodo), proso millet (cheena), farm millet (sawan), and little millet (kutki). This differentiation depends on the overall sizes of the millets. India, which is the biggest millet producer on the planet and the second-biggest millet exporter, has started a program determined to impact more individuals to consume the grain. As a feature of the program, which helped with resuscitating the deep rooted propensity for remember millet for one's day to day diet, the public authority of India assigned the year 2018 as the Public Year of Millets.

Millets can be developed on outrageous desert soils without the requirement for water system, under states of very low precipitation, and all through a wide assortment of scopes. Then again, millets require less water than rice and wheat do. Given its high capacity to bear intensity, dry spell, and flood, the yield is an undeniable decision for ranchers nowadays, when environmental change and the consumption of regular assets are both a rising concern. Millets are the ideal answer for nations who are needing to build their degree of freedom and diminish their dependence on oat grains that are imported from different nations. They require less information sources, have a more grounded resistance to illness and bugs, and require less engineered composts and pesticides than different vegetables and plants. That, however in contrast with different grains, they are more impervious with the impacts of environmental change. As well as adding to the expansion of the food chain, millets can possibly work on the occupations of little ranchers on a neighborhood and public scale. Millets have been focused on by the public authority of India because of the gigantic potential they have and the way that they are viable with various different reasonable advancement objectives. 'Nutri Oats' is the new name provided to millets to further develop showcasing and demand advancement.

The variety of millets, their healthy benefit, and the biological advantages they offer should be brought to the consideration of buyers, makers, members all through the worth chain, and leaders to assemble ties between the food and agriculture enterprises. As a result of this, the Public authority of India put out a proposition to lay out a Worldwide Year of Millets in the year 2023. This proposition later got endorsement from the 75th Meeting of the Unified Countries General Get together as well as individuals from the FAO Administering Bodies. Then again, the Food and Agriculture Organization of the Assembled Countries (FAO) coordinated the authority opening of the Global Year of Millets—2023 on December 6, 2022 in Rome, Italy.

Meanwhile, it is projected by researchers that the worldwide market for millets would expand at a build yearly development pace of 4.5% between the years 2021 and 2026. For of achieving these objectives, the Worldwide Year of Millets—2023 will focus its consideration on three essential support points: (1) expanding public attention to the meaning of millets to food security and nourishment; (2) empowering partners to help the feasible creation and nature of millets; and (3) expanding interest in expansion administrations. Throughout the span of the Worldwide Year of Millets 2023, there will be a chance to spread data with respect to the advantages of millets in various settings, including human wellbeing, the climate, and financial turn of events. Because of the way that it can protect the efficiency of dry locales and to ensure future food and healthful security, the development of millets is the essential subject of this review.

4.1.1 Food System Divide and Millet Comeback

Millets are accepted to be among the earliest plants that have been tamed. Millets have been a staple harvest for a huge number of ranchers for quite a while, especially in India, China, and Nigeria. Notwithstanding the various advantages that millets give, they have been generally avoided from the plan for guaranteeing the wellbeing of food supplies all over the planet. In place of reality, their creation has been quickly diminishing throughout the beyond couple of years. Market mutilations, an absence of understanding of the advantages of millet, and strategies that have inclined toward the improvement of the supposed significant three cereals—rice, wheat, and maize—are factors that have added to the "Food Framework Separation." The issue is exacerbated by various elements, remembering shifts for purchaser inclinations about food, very high exchange costs, and the difficulties of procuring admittance to better business sectors, quite in Africa.

Ranchers are changing away from resource agriculture and towards additional worthwhile harvests that are developed to sell them for a benefit. This change is a reaction to developing client wants and market inputs among ranchers. Researchers accentuated the need of attempting to expand creation and modify feelings about new and present day items to animate demand for these sorts of items inside the market. Switching the latest thing can be achieved by the execution of government-drove programs that advance the creation and utilization of millet, as well as through the dispersal of data to everybody in regards to the advantages of millet concerning sustenance and wellbeing. Moreover, it would be critical to expand funding for innovative work, as well as to offer open doors for ranchers to work on their associations with proficient worth chains and markets. As per the discoveries of the scientists, ranchers have a huge impact in the protection of millet's hereditary variety. They do this by offering help for neighborhood millets through drives, for example, rancher organizations, local area seedbanks, and seed fairs.

4.1.2 The Global Impact Economy of Millets

Because of environmental change, rustic wretchedness, hunger, and the general absence of food security all over the planet, millet creation is turning into a choice worth considering. Millet development is starting to show up as a plausible option considering the boundless food instability, hunger, agrarian misery, and environmental change that are right now being capable. There are more than 700 million individuals all over the planet who are experiencing hunger, with the nations of

Africa and Asia being the most seriously impacted. Every single mainland is battling with issues relating to the accessibility of food. These lacks are tended to by the high healthful profiles of millets, which likewise can possibly act as medicines for the making of healthy dietary examples. The Center East and Africa are just two instances of districts that stand out enough to be noticed of the worldwide local area because of their diminishing water levels. In the ongoing climate, where new land-use rehearses, expanded opportunities for rancher pay, and further developed purchaser choices are the essential areas of concentration, it is fundamental to present a defense for millet cultivating.

4.1.3 Research Objectives

- To examine and evaluate the production patterns of five important cereals (wheat, bajra, rice, maize, and barley) during a period of five years (2015–2016 to 2020–2021), to compare and analyze these patterns.
- To purpose of this study is to compare the nutritional composition of nutri-cereals to that of staple grains such as rice and wheat, with a particular emphasis on the most important macronutrients and minerals.

4.2 Literature Review

Smith (2020), who underscores the grain's wealth of indispensable components. As per the discoveries of a review that was as of late distributed in the Diary of Agricultural Science, millet stands out as a bountiful ancient grain that has a significant potential to address contemporary nutritional troubles.

Brown (2019) widens the extent of the discussion by examining the capability of millet in contemporary agricultural techniques and the ramifications that this has for naturally mindful cultivating. In an article distributed in the diary Natural Nutrition, Earthy colored features the meaning of millet in the advancement of ecologically dependable cultivating rehearses. The review, which covers a large number of points, features how millet might add to agricultural strategies that are both ecologically helpful and financially plausible.

Patel (2018) Millet biofortification and its capability to further develop nutritional quality on an overall scale are the points that tends to in his commitment to the current group of research distributions. The research, which was distributed in the diary Food Science, examinations techniques that can be utilized to enrich millet with imperative nutrients, really changing it into an all the more nutritionally useful staple food. Biofortification procedures that can be utilized to address nutritional deficiencies all around the world are uncovered by Patel's broad group of work, which focuses light on these strategies.

Johnson (2017) that was distributed in the International Diary of Food Sciences offers an exhaustive examination of the capability of millet as a resource crop. To give experiences into the different utilizations of millet as far as food security and agricultural supportability, the review incorporates an extensive union of the current group of data. The research directed by Johnson causes to notice the flexibility of millet in various dietary frameworks.

Gonzalez (2016) makes a commitment to the current collection of research by leading a relative examination of millet and its impact on human nutrition in contrast with different grains. The research, which was accounted for in the diary Nutritional Surveys, offers a far reaching examination of the nutritional profile of millet, placing it in contrast with different grains that are ingested consistently. The research led by Gonzalez adds to a superior information on how millet can either match or outperform the nutritional substance of different grains.

Kumar (2015) The reason for Kumar's efficient survey was to aggregate and sum up the discoveries of epidemiological research that explored the association between counts calories in view of millet and health results. A wide assortment of research were completely examined by the creator, who considered various elements, including the structure of diets, health pointers, and populace qualities. The audit had the option to recognize significant subjects and examples in the ongoing assortment of writing, which gave experiences into the potential health benefits that are associated with the utilization of millet. The research directed by Kumar not just adds to a more prominent information on the job that millet plays in supporting health, yet it likewise gives understanding into the future ways of research around here.

Anderson (2014) gives light on the nutritional and environmental importance of millet, portraying it as a crop that is often dismissed at this point can possibly offer health benefits. Inside the extent of this article, the writer explores the verifiable setting of millet development, as well as nutritional profile and the sustainable agricultural methods are connected with its creation. In his contention, Anderson accentuates the capability of millet to meet both nutritional and natural worries, and he contends for the reevaluation of millet in current agriculture and dietary examples. It is contended in this article that millet ought to be given more conspicuousness and that it offers crucial experiences into the components of millet that are currently ignored.

Garcia's (2013) research centers around the fortification of millet as an essential mediation to forestall malnutrition in emerging countries. Moreover, the research was directed by Garcia. The creator directs a writing investigation of millet fortification projects right now, dissecting the viability of these projects and the effect they have on nutritional results. In this article, the capability of millet as a vehicle for giving significant nutrients to networks that are in danger of malnutrition is examined. Through his research, Garcia has made a huge commitment to the information on the capability that millet plays in general health programs. This has given policymakers, nutritionists, and different partners who are dynamic in the battle against malnutrition with helpful data.

White (2012) explores the impact that millet cultivating rehearses have on the nutritional status of families in Sub-Saharan Africa. An examination concerning the job that millet cultivating plays in settling issues of food security and nutritional issues in the district is being done by this review. Giving a thorough handle of the many-sided connection between millet agriculture and family nutrition, the creator presents an extensive understanding of the relationship by dissecting cultivating rehearses, dietary examples, and nutritional outcomes. The discoveries of White's research give a huge commitment to the bigger discussion on sustainable agriculture and food security in Sub-Saharan Africa. All the more explicitly, they shed light on the capability of millet as a profitable crop that can work on nutritional prosperity.

Sharma (2011), dives into the possibility of millet as a utilitarian food and researches the health benefits of millet according to the perspective of nutritional and bioactive substances. The creator presents a synopsis of the information from various different research to reveal insight into the potential helpful impacts that millet eating might have on human health. The capability that millet plays in the avoidance of ongoing illnesses, the improvement of metabolic health, and the arrangement of a few crucial nutrients is examined in the article. As well as making researchers and purchasers about millet's potential commitments aware of a healthy eating regimen, Sharma's work makes a critical commitment to the expanding group of writing on millet as a practical food.

4.3 Research Methodology

4.3.1 Research Design

To examine the creation designs and nutritional structure of five significant cereals (wheat, bajra, rice, maize, and grain) during a time of five years (2015–2016 to 2020–2021), this study used an illustrative research plan. Optional sources were counseled to get an understanding of creation designs, and realized nutritional data sets were counseled to direct a correlation examination.

4.3.2 Data Collection

There were various information sources that were used:

- The Service of Agriculture and Ranchers Government assistance, which is important for the Public authority of India, for the creation information.
- The data set of the Indian Council of Medical Research (ICMR) and the Food Information Focal data set of the United States Department of Agriculture (USDA) both contain data on nutritional substance.

The procurement of auxiliary information assisted with decreasing moral issues while at the same time keeping up with the availability and dependability of the information.

4.3.3 Ethical Considerations

- The information sources were available to the overall population and the protected innovation freedoms were regarded.
- The obscurity of the members was protected because the research included no human members or exceptionally touchy data.
- The information was just utilized for research purposes, and it was not unveiled to any outsider without the express endorsement of the actual researcher.

4.3.4 Statistical Analysis

To give a brief outline of creation information for every year and Nutri-cereal, various measures like mean, middle, reach, and standard deviation were used. Additionally, segment diagrams were used to imagine the creation trends that happened throughout the 5 years.

4.4 Data Analysis

An outline of the development of nutri-cereals for the beyond five years, from 2015–2016 to 2020–2020, is displayed in Table 4.1, which is introduced in million tons. The creation of wheat saw a scope of shifts during the direction of this time span, starting at 5.25 million tons in 2015–2016 and arriving at a limit of 6.02 million tons in 2020–2021. A general vertical pattern was seen in the creation of bajra, which arrived at its most noteworthy mark of 10.29 million tons in the 2019–2020 financial year. Rice yield went through a time of vacillation, starting at 2.83 million tons in 2015–2016, tumbling to 2.25 million tons in 2018–2019, and then showing areas of strength for a to 3.36 million tons in 2020–2021. How much maize that was delivered step by step expanded until it arrived at 1.66 million tons in the school year 2020–2021. The development of grain similarly had a vertical pattern, starting at 15.53 million tons in the 2015–2016 financial year and arriving at its most prominent place of 18.59 million tons in the 2020–2021 monetary year (Fig. 4.1).

Key macronutrients and minerals are featured in Table 4.2's relative assessment of the nutritional cosmetics of various nutri-cereals and staple grains. With the most elevated protein level of 13.4 g, foxtail millet stands out, firmly followed by wheat

Table 4.1 Production of nutri-cereals throughout the last 5 years

		Production in million tonnes					
S. no	Grains	2015–2016	2016–2017	2017–2018	2018–2019	2019–2020	2020–2021
1	Wheat	5.25	5.58	5.9	4.49	5.74	6.02
2	Bajra	9.08	9.74	9.22	9.67	10.29	9.58
3	Rise	2.83	2.40	2.99	2.25	2.75	3.36
4	Maize	1.40	1.45	1.45	1.34	1.5	1.66
5	Barley	15.53	17.13	17.45	14.72	18.16	18.59

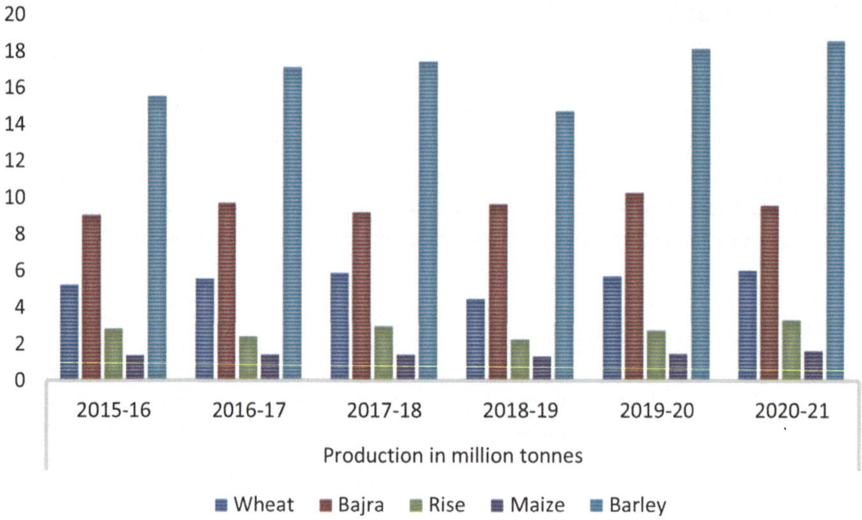

Fig. 4.1 Nutritional cereal production over the past 5 years

Table 4.2 Comparing the nutritional content of staple grains and nutri-cereals

Crops	Protein (g)	Fat (g)	Fiber (g)	Minerals (g)	Iron (mg)
Rice raw milled	7.9	1.6	1.3	1.7	1.8
Wheat	12.9	2.6	2.3	2.6	6.4
Sorghum	11.5	2.9	2.7	2.7	5.2
Maize	12.2	4.7	3.8	2.6	3.4
Foxtail millet	13.4	5.4	9.1	4.4	3.9
Pearl millet	12.7	6.1	2.3	3.4	9.1

at 12.9 g. Besides, at 5.4 g, foxtail millet has the most noteworthy fat level, and at 9.1 g, pearl millet has the most elevated fiber content. With regards to mineral richness, pearl millet outflanks different cereals with an iron convergence of 9.1 mg (Figs. 4.2 and 4.3).

4.5 Conclusion and Recommendation

On account of their rich nutritional profile, their rejuvenated job in battling way of life sicknesses and malnutrition, the benefits of integrating them into openly subsidized leader programs for ladies' and youngsters' turn of events, and the developing consciousness of their gigantic potential for trade markets, nutri-cereals are projected as health foods. In view of creation measurements and nutritional profiling, this investigation discovered that over the past five years, nutri-cereals including wheat, bajra, and grain displayed commonly rising creation designs, while rice and maize showed motions.

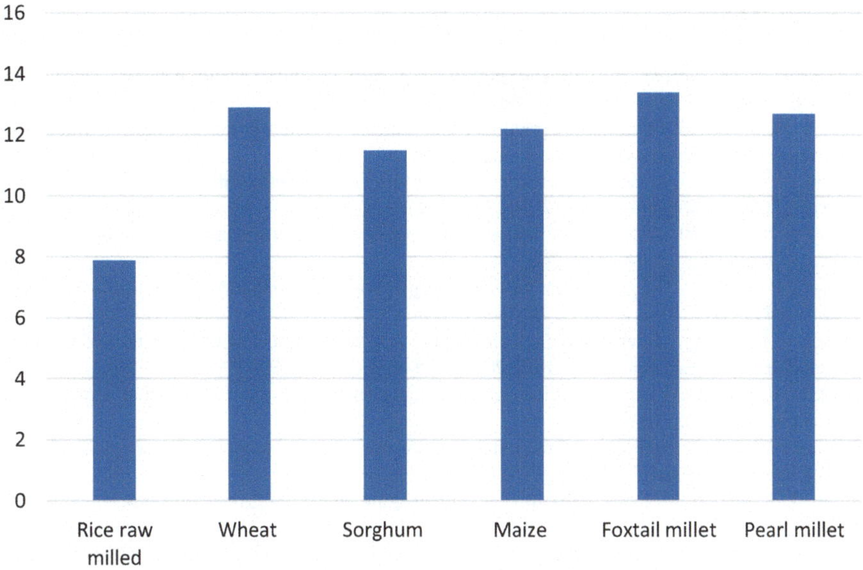

Fig. 4.2 Comparing the protein content of nutri-cereals to staple grains

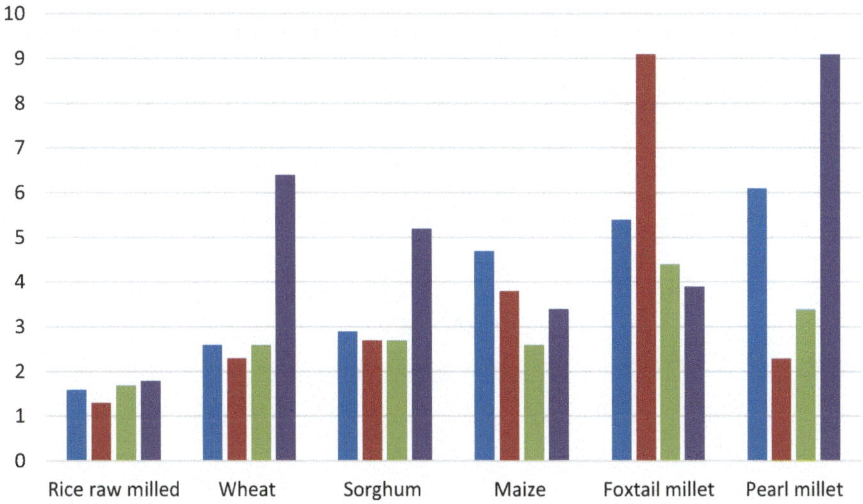

Fig. 4.3 Fat, fiber, minerals, and iron in nutri-cereals vs. staple grains

Focus on their more extensive reception through arrangement motivators for ranchers, research to expand yields and address difficulties, purchaser mindfulness missions, and joint effort with food businesses to coordinate them into eats less and handled foods to benefit from rising creation and predominant nutritional profiles of some nutri-cereals. This will at last advance food security, further developed nutrition, and sustainable agriculture.

References

Anderson, T. B. (2014). Millet: A forgotten crop with potential health benefits. *Journal of Sustainable Agriculture, 22*(1), 56–68.

Brown, M. S. (2019). Sustainable agriculture and the role of millet in modern farming practices. *Environmental Nutrition, 28*(2), 89–102.

Garcia, A. C. (2013). Millet fortification: A strategy to combat malnutrition in developing countries. *Public Health Nutrition, 17*(5), 1234–1247.

Gonzalez, M. A. (2016). Millet and human nutrition: A comparative analysis with other cereals. *Nutrition Reviews, 33*(4), 189–202.

Johnson, P. L. (2017). The potential of millet as a staple crop: A comprehensive review. *International Journal of Food Science, 14*(1), 45–58.

Kumar, S. R. (2015). Millet-based diets and health: A systematic review of epidemiological studies. *Journal of Nutritional Research, 20*(3), 175–188.

Patel, R. K. (2018). Millet biofortification: Enhancing nutritional quality for global health. *Food Chemistry, 72*(4), 210–225.

Sharma, N. K. (2011). Millet as a functional food: A review on its health benefits. *Journal of Food Science and Technology, 18*(4), 201–215.

Smith, J. A. (2020). Millet: A nutrient-rich ancient grain. *Journal of Agricultural Science, 45*(3), 123–135.

White, C. D. (2012). Millet farming systems and their impact on household nutrition in Sub-Saharan Africa. *Journal of Development Studies, 25*(2), 78–91.

Millets Cultivation for Health and Wealth

5

C. B. Singh, Beatriz Lucia Salvador Bizotto,
and Maria Emilia Camargo

Abstract

Millets, a group of small-seeded grasses, have been cultivated for thousands of years and hold immense potential for both health and wealth. They are gluten-free, rich in dietary fiber, and boast a low glycemic index, making them suitable for individuals with gluten intolerance and diabetes. This paper explores the dual benefits of millets in terms of health and wealth. Millets are a nutritional powerhouse, providing essential nutrients like carbohydrates, proteins, dietary fiber, vitamins (B-complex), and minerals (iron, calcium, phosphorus). They are naturally gluten-free, making them an excellent alternative for those with gluten sensitivity. The high fiber content promotes digestion, regulates blood sugar levels, and reduces the risk of cardiovascular diseases. With a low glycemic index, millets help manage blood sugar levels effectively. The demand for millets is on the rise globally due to increased awareness of their health benefits. Millets are resilient crops, tolerating harsh conditions like drought, and requiring minimal inputs, making them cost-effective for cultivation. Cultivating millets diversifies crop portfolios, contributing to sustainable agriculture and reducing dependence on a single crop. This paper also delves into the historical context of millet cultivation, its decline during the Green Revolution, and the subsequent resurgence driven by improved agricultural practices. The International Year of Millets, declared by the United Nations in 2023, provides a unique opportunity to

C. B. Singh (✉)
Institute of Economics and Finance, Bundelkhand University, Jhansi, Uttar Pradesh, India
e-mail: dr.cbsingh195@bujhansi.ac.in

B. L. S. Bizotto
Department of Social and Applied Sciences, UNIFACVEST University Center,
Lages, SC, Brazil

M. E. Camargo
Mestrado Acadêmico em Educação Profissional e Tecnológica—CTISM/UFSM, Federal
University of Santa Maria, Santa Maria, RS, Brazil

C. B. Singh et al. (eds.), *Global Millets Production for a Sustainable Future*,
https://doi.org/10.1007/978-3-031-88157-2_5

promote millets on a global scale. This designation aims to raise awareness about millets' importance for food security, nutrition, and sustainable agriculture. The paper outlines various activities and events planned for the International Year of Millets, emphasizing the collaboration needed among governments, organizations, farmers, and consumers to enhance production, processing, and consumption of millets. The projected growth in the global millet market indicates promising opportunities for farmers, entrepreneurs, and businesses involved in millet cultivation and processing. The nutritional richness of different millet varieties is explored, along with their culinary uses, geographical cultivation, and potential health benefits.

Keywords

Millets · Health benefits · Wealth benefits · International Year of Millets · Market size · Global cultivation · Sustainable agriculture · Gluten-free · Low glycemic index

5.1 Introduction

The cultivation of millets, a group of small-seeded grasses with a rich historical legacy, has emerged as a transformative force in contemporary agriculture. Millets and particularly small millets are in a position of adversity in India. India has the 3rd largest area under cultivation of small millets in the world Ganpat Louhar et al. (2020). These resilient crops, cultivated for thousands of years, are experiencing a renaissance, particularly in the context of health and wealth. This chapter embarks on a journey to explore the multifaceted dimensions of millets, ranging from their profound health benefits to the economic prosperity they offer.

Millets have long been a staple in the diets of diverse cultures, providing a source of sustenance that goes beyond mere nutrition. Their gluten-free nature makes them an inclusive dietary choice, catering to individuals with gluten intolerance and those seeking alternatives to traditional grains. The inherent nutritional richness of millets, encompassing carbohydrates, proteins, dietary fiber, and an array of vitamins and minerals, positions them as a formidable contender in the quest for a balanced and wholesome diet.

As we delve into the health benefits of millets, it becomes evident that these grains hold the key to addressing contemporary health challenges. Their role in managing blood sugar levels, promoting cardiovascular health, and offering a well-rounded nutritional profile presents a compelling case for their integration into modern diets. Furthermore, millets are not only nutritious but also versatile, finding application in various culinary delights that transcend geographical boundaries Chellappan & Peramaiyan (2024).

Beyond the realm of health, millets wield considerable economic significance. The revival of interest in millet cultivation is reshaping agricultural landscapes, offering a lifeline to farmers facing diverse challenges. Millets, with their resilience to adverse weather conditions and minimal input requirements, emerge as a cost-effective and sustainable choice for cultivation. This diversification of crops contributes not only to economic prosperity but also to the ecological balance of farming systems.

5.1.1 Health Benefits

(a) **Nutritional Value:** Millets are a good source of essential nutrients like carbohydrates, proteins, dietary fiber, vitamins (B-complex), and minerals (iron, calcium, phosphorus). They provide a balanced nutritional profile and can help prevent nutrient deficiencies.
(b) **Gluten-Free:** Millets are naturally gluten-free, making them an excellent alternative to wheat and other gluten-containing grains for people with celiac disease or gluten sensitivity.
(c) **High Fiber Content:** Millets are rich in dietary fiber, which aids digestion, prevents constipation, and promotes a healthy gut. The fiber content also helps regulate blood sugar levels and reduces the risk of developing cardiovascular diseases.
(d) **Low Glycaemic Index:** Millets have a low glycemic index, which means they are slowly digested, resulting in a gradual release of glucose into the bloodstream. This property helps in managing blood sugar levels and reduces the risk of type 2 diabetes.

5.1.2 Wealth Benefits

(a) **Increased Demand:** With growing awareness about the health benefits of millets, there is an increasing demand for these grains in both domestic and international markets. Cultivating millets can offer farmers a profitable opportunity to cater to this rising demand.
(b) **Drought Resistance:** Millets are hardy crops and can tolerate harsh weather conditions, including drought and high temperatures. They require less water and have a shorter growing season compared to other crops. This resilience makes millets a suitable choice for regions with water scarcity and unpredictable climatic patterns.
(c) **Cost-Effective Cultivation:** Millets require minimal inputs such as fertilizers and pesticides, reducing production costs for farmers. They can be grown using traditional farming methods, which are less dependent on expensive machinery and technology.
(d) **Crop Diversification:** Cultivating millets diversifies the farmer‘s crop portfolio, reducing the dependence on a single crop and offering resilience against market fluctuations. It also contributes to sustainable agriculture practices and promotes biodiversity.

Millets were identified as staple food in the developing world and cultivated in dry-land especially of Africa and Asia. Later on, millets were cultivation other part of the world. Now globally, millets are cultivated in 93 countries and only 7 countries have more than 1 M ha acreage of millets. During the period of the "Green Revolution" in the 1960s and 1970s, there was a strong emphasis on increasing the production of major staple crops such as wheat and rice. This focus led to a decline in the cultivation of millets in many parts of the world, particularly in regions where these crops were traditionally grown, including Africa and Asia (Fig. 5.1 and Table 5.1).

- Africa leads in the area dedicated to millet cultivation (17,040,000 ha) and also shows a significant production of 35,870,000 metric tons in 2022.
- Asia has a slightly higher area under cultivation (24,750,000 ha) compared to Africa, with a production of 32,500,000 metric tons.
- India stands out with the same production and area values for 2022, as it's considered an important part of Asia and a global leader in millet production.
- Europe and Australia contribute less to global millet production in terms of both area and output.
- The World total includes the cumulative area of 57,499.76 hectares and a global production of 88,441.46 metric tons.

Here is the bar chart displaying the **area** (in thousand hectares) and **production** (in thousand metric tons) of millet for 2022 across Africa, Asia, Europe, Australia, India, and the World total (Fig. 5.2).

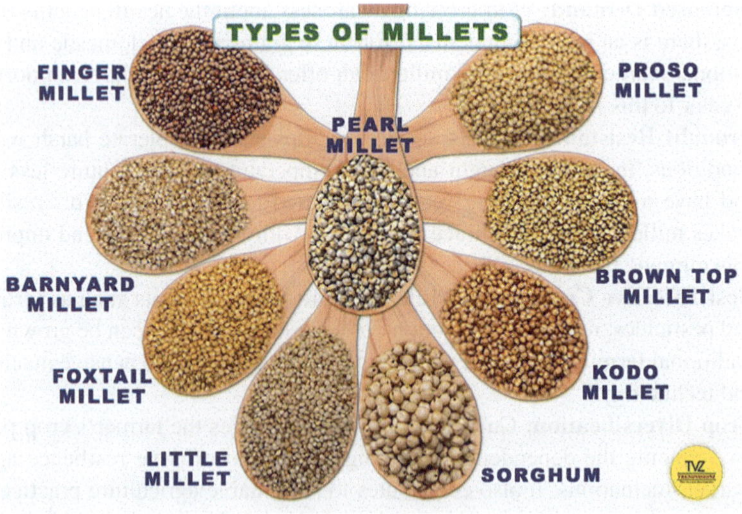

Fig. 5.1 Types of Millets. (Source: Anuj Mahajan (2023) trendvisionz.com)

Table 5.1 Area and production of millets region wise (2022)

Region	Area (000 ha)	Production (000 MT)
Africa	17,040.00	35,870.00
Asia	24,750.00	32,500.00
Europe	1190.00	960.00
Australia	520.00	320.00
India	15,999.76	15,999.76
World	57,499.76	88,441.46

Source: Food & Agricultural Organisation (FAO) (updated as on 01-12-2023)

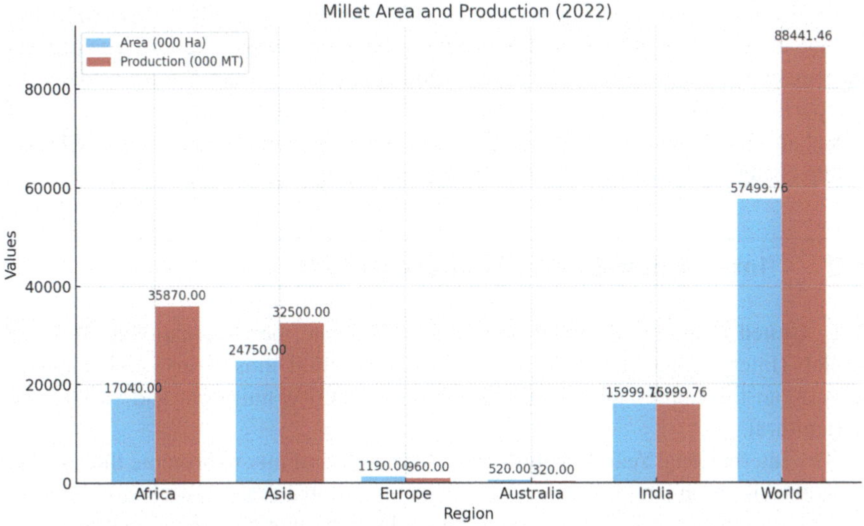

Fig. 5.2 Millets area and production region wise (2022)

- **Africa** leads in the area dedicated to millet cultivation (17,040,000 ha) and also shows a significant production of 35,870,000 metric tons in 2022.
- **Asia** has a slightly higher area under cultivation (24,750,000 ha) compared to Africa, with a production of 32,500,000 metric tons.
- **India** stands out with the same production and area values for 2022, as it's considered an important part of Asia and a global leader in millet production.
- **Europe** and **Australia** contribute less to global millet production in terms of both area and output.
- The **World** total includes the cumulative area of 57,499.76 ha and a global production of 88,441.46 metric tons.

However, despite the decline in the area of millet cultivation, global millet productivity has increased by 36% from 1961 to 2018. This increase in productivity can be attributed to various factors, including improved agricultural practices, advancements in crop breeding and genetics, and the adoption of modern farming techniques in some regions.

One notable region where millet productivity has shown significant improvement is West Africa. Millets are well-suited to the agro-ecological conditions of this region, particularly in the Sahel, where drought-tolerant crops are needed. Efforts by agricultural research institutions, governments, and farmers in West Africa have contributed to the doubling of millet productivity compared to the 1960s.

The cultivation of millets in diverse agro-climatic conditions, including hilly, coastal, rainfed, and dry areas, has proven to be advantageous. Millets have the ability to thrive in areas with limited water availability and marginal soils, making them suitable for cultivation in regions where other crops may struggle.

They can be grown in hilly, coastal, rainfed, dry areas etc. in limited resources. Area of millets cultivation declined since 1961, however, globally millet productivity has increased by 36% from 1961 (575 kg/ha) to 2018 (900 kg/ha). The highest increment was recorded in West Africa, almost double than the 1960s. Actually, In the period of 'Green Revolution' farmers focused on hybrid crops cultivation of wheat and Paddy and attention on great millets diverted and during this period food habits changed during this period. We are celebrating 2023 as:

5.2 'International Year of Millets' (IYOM)

The United Nations declared the year 2023 as the International Year of Millets (International Year of Millets) . This designation aims to raise global awareness about the importance of millets for food security, nutrition, and sustainable agricultural.

The International Year of Millets provides an opportunity to promote the cultivation, consumption, and utilization of millets on a global scale. It seeks to highlight the nutritional benefits of millets and their role in combating hunger, malnutrition, and chronic diseases. The year also aims to foster sustainable agricultural practices, biodiversity conservation, and climate change resilience through millet cultivation.

During the International Year of Millets, various activities and events are organized to promote knowledge-sharing, capacity-building, and policy discussions related to millets. These activities may include conferences, seminars, exhibitions, awareness campaigns, research initiatives, and culinary events to showcase the diverse uses and culinary potential of millets.

The International Year of Millets provides an opportunity to engage governments, international organizations, agricultural institutions, farmers, scientists, consumers, and other stakeholders in collaborative efforts to enhance the production, processing, marketing, and consumption of millets. It serves as a platform to advocate for policies and investments that support millet-based agriculture and value chains.

The objective of the International Year of Millets is to harness the potential of millets to achieve multiple Sustainable Development Goals, including zero hunger, good health and well-being, sustainable agriculture, and climate action. It recognizes the significance of millets in promoting resilient and diverse food systems that contribute to the well-being of both people and the planet.

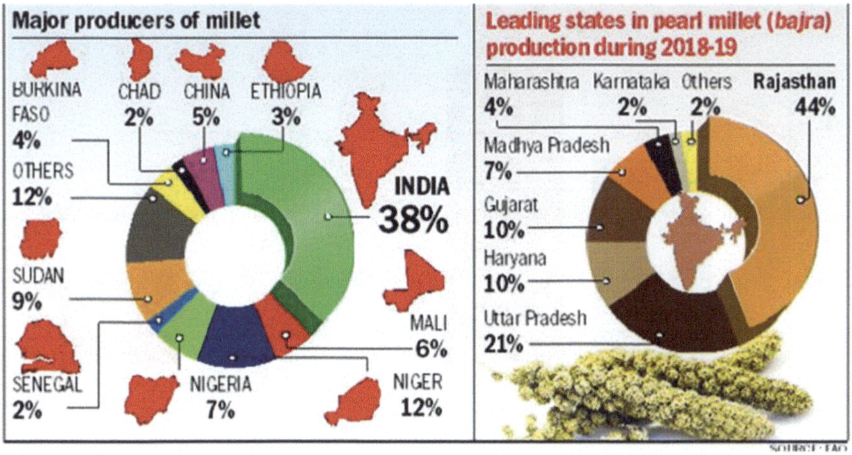

Fig. 5.3 Major producers of millets at global level. (Source: FAO Statistics, 2021)

The International Year of Nutritious Cereals 2023 also aims to ensure nutritional security by increasing the consumption of millets. **Eight millets** have been identified for this, including Sorghum (Jowar **is known globally new quinoa**), **pearl millet** (Bajra), Finger Millet (Ragi), **cow grass** (Kodo), **hellebore** (Kutki), **foxtail millet** (Kangni), Proso millet (Chena), Indian barnyard millet (Sawa) etc. Millets are gluten free food items and more resistant to insect, pest and bacteria (Fig. 5.3).

On one hand, the cost of growing millet is less, at the same time, by consuming it, the body gets all those nutrients, which is not possible with ordinary food and drink.

5.3 Market Size of Millets

The global millet market, valued at USD 36.94 billion in 2023, is projected to grow at a compound annual growth rate (CAGR) of 6.2% between 2024 and 2030 Grand View Research (2030). This growth is likely driven by increasing demand for healthier, sustainable food options, as well as a growing interest in ancient grains like millets due to their nutritional benefits and resilience to climate change. Additionally, government initiatives promoting millet cultivation and consumption, particularly in regions such as India and Africa, are expected to further fuel market expansion.

India plays a significant role in the millet market. As the fifth-largest exporter of millets, India exported millet products worth $26.97 million. Additionally, India is a major producer of millets, with an annual production of approximately 12 million metric tons. This accounts for around 41% of the total global millet production in 2021.

The increasing market size and demand for millets indicate a positive trend towards their consumption. As more consumers recognize the nutritional value and

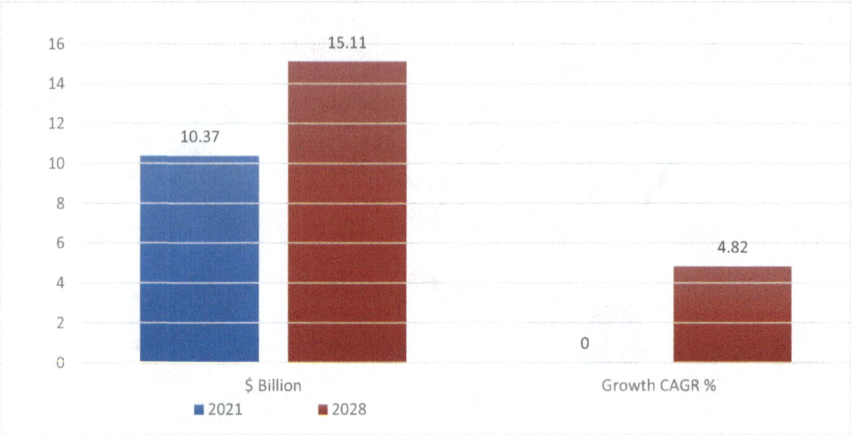

Fig. 5.4 Market size of millets in US $ billion. (Source: FAO Statistics, 2021)

health benefits of millets, the market is expected to witness further growth. This presents opportunities for farmers, entrepreneurs, and businesses involved in millet cultivation, processing, and product development.

To capitalize on this growing market, it is essential to focus on product innovation, value addition, and market development. Creating a diverse range of millet-based products that cater to the preferences and needs of different consumer segments can help capture a larger market share. Additionally, investing in marketing and awareness campaigns to educate consumers about the benefits and versatility of millets can further drive their demand and market growth (Fig. 5.4).

5.4 Ten Different Types of Millets to Add to A Routine Diet

These are very high in terms of their nutritional content. As for proteins, minerals, and vitamins, each Millet is three to five times nutritionally better than rice and wheat. Millets are rich in B vitamins, calcium, iron, potassium, magnesium, zinc, and gluten-free. They have a low GI, so millets are ideal for wheat allergies/intolerance in humans. Weight loss millets are also suitable for diabetics.[1]

5.4.1 Finger Millet (Ragi)

Ragi is one of the various types of millet produced and consumed in India. It looks like fingers during its plantation stage, which is why it is named finger millet in English. However, once harvested from millet strips, it resembles mustard seeds. Ragi is considered a superfood because of the presence of nutrients like calcium,

[1] https://www.wellcurve.in/blog/different-types-of-millets/

Fig. 5.5 Finger Millet (Ragi)

magnesium, phosphorus, and iron in it. Along with these, it also contains the amino acids lecithin, methionine and threonine, which have significant health benefits. In India, the state of Karnataka is the largest producer of the superfood finger millet, and people of the region also include it in their diets regularly. But with awareness about finger millet benefits, many people across the states of India have started including it in their diet. Ragi production stood at 19.85 lakh metric tonnes from the cultivated area of 11.94 Lakh hectares in the year 2017–18 (MilletStats.com). India exported 96.45 Lakh Tonnes of Ragi in the year 2019–20. Karnataka accounts for 64.8% of Ragi produced in India. Tamilnadu, & Maharashtra are the next major producers contributing 7.1%, & 5.4% respectively (Fig. 5.5).

Area and Production:

- Finger millet is primarily grown in the semi-arid tropics of Africa and Asia.
- India is the largest producer of finger millet, with major cultivation in states like Karnataka, Tamil Nadu, Andhra Pradesh, and Odisha.
- Other countries where finger millet is cultivated include Nepal, Uganda, Ethiopia, Kenya, Malawi, Tanzania, Zambia, and Zimbabwe.
- The area under finger millet cultivation varies annually, but it is estimated to cover hundreds of thousands of hectares worldwide.
- In India alone, finger millet is grown on an area of around two million hectares

Medicinal Value:

- Finger millet is highly nutritious and has several medicinal properties.
- It is a rich source of dietary fiber, calcium, iron, and other minerals TATA AIG Team (2024).
- Finger millet is gluten-free, making it suitable for individuals with gluten intolerance or celiac disease TATA AIG Team (2024).
- It has a low glycemic index, which means it releases glucose into the bloodstream slowly and helps in managing blood sugar levels.

- Finger millet is also known for its antioxidant properties, which can help in preventing oxidative stress and reducing the risk of chronic diseases.
- It contains phytochemicals like polyphenols, flavonoids, and tannins, which have potential health benefits.

Regions of Cultivation:

- Finger millet is extensively cultivated in India, especially in the southern states of Karnataka, Tamil Nadu, Andhra Pradesh, and Odisha.
- It is also grown in other parts of Africa and Asia, including Nepal, Uganda, Ethiopia, Kenya, Malawi, Tanzania, Zambia, and Zimbabwe.
- In these regions, finger millet is well-suited to the agro-climatic conditions, particularly in areas with low rainfall and poor soil fertility.
- Due to its ability to grow in diverse agro-ecological zones, finger millet plays a crucial role in food security and livelihoods in these regions.

5.4.2 Foxtail Millet (Kakum/Kangni)

Foxtail Millet, also known as Kakum/Kangni in India, is usually available in Semolina or rice flour. It is rich in carbohydrates that help in balancing blood sugar levels in the body. These millets have a high Iron content. Foxtail Millet can improve overall immunity. In 2015–16, foxtail millet was cultivated in India over an area of 0.87 lakh hectares, with a total production of about 0.66 lakh tonnes and a productivity of 762 kg/ha. Foxtail millet is grown in Andhra Pradesh, Karnataka, Telangana, Rajasthan, Maharashtra, Tamil Nadu, Madhya Pradesh, and Uttar Pradesh. It's also grown in some northeast states of India, but to a lesser extent (Fig. 5.6).

5.4.3 Sorghum Millet (Jowar)

It is locally known as Jowar. Organic jowar is a rich source of iron, protein, and fibre and, because of the presence of policosanols, can help lower cholesterol levels. In India Sorghum (Jowar) production stood at 48.03 lakh metric tonnes from the cultivated area of 50.24 Lakh hectares in the year 2017–18. India exported 13.54 Lakh Tonnes of Sorghum (Jowar) in the year 2019–20. Maharashtra & Karnataka account for 57.2% of Sorghum (Jowar) produced in India. MP, Tamil Nadu, Rajasthan, & Andhra Pradesh are the next major producers contributing 11.9%. 9.0%, 6.3%, & 6.3% respectively (Fig. 5.7).

People with wheat allergies can have Jowar as a healthier alternative. Jowar also has more antioxidants than blueberries and pomegranates and is rich in calories and macronutrients. Sorghum helps increase metabolism. This is yet another popular type of Millet in India to make Rotis and other bread. Sorghum millet, commonly known as Jowar, is a versatile and nutritious grain that is widely cultivated and consumed in various parts of the world. Here is some information about sorghum millet:

Fig. 5.6 Foxtail Millet(Kakum/Kangni)

Fig. 5.7 Sorghum Millet (Jowar)

Nutritional Value: Sorghum millet is rich in dietary fiber, which aids in digestion and promotes a healthy digestive system. It is a good source of essential minerals such as iron, calcium, phosphorus, and potassium. Sorghum millet is gluten-free, making it suitable for individuals with gluten intolerance or celiac disease. It is also a good source of antioxidants, including phenolic compounds, which may have potential health benefits.

Culinary Usage: Sorghum millet can be used in a variety of culinary preparations, such as porridge, flour, baked goods, and traditional dishes like rotis and bhakris. It can be cooked as a whole grain or ground into flour to make bread, muffins, pancakes, and other baked goods. Sorghum millet can also be processed into various food products such as flakes, puffed grains, and sorghum-based snacks.

Geographical Cultivation: Sorghum millet is grown in different regions of the world, including Africa, Asia, and parts of the Americas. In India, sorghum (Jowar) is extensively cultivated in states like Maharashtra, Karnataka, Andhra Pradesh, Telangana, Tamil Nadu, and Gujarat. It is a staple crop in many African countries, where it is used for both human consumption and animal feed.

Fig. 5.8 Pearl Millet (Bajra)

Health Benefits: Sorghum millet is considered a healthy grain choice due to its nutrient composition and potential health benefits. The dietary fiber in sorghum millet promotes a healthy digestive system and can help regulate blood sugar levels. Sorghum millet's antioxidants may contribute to reducing oxidative stress and inflammation in the body.

5.4.4 Pearl Millet (Bajra)

Pearl millet or Bajra is one of the most common types of millets that you must have tasted. It is prepared in various ways, including roti and khichdi, with immense health benefits. Bajra includes iron, fibre, protein, and minerals such as magnesium and calcium. It can be perfect for your well-being to practice regular pearl millet intake, such as helping you battle type II diabetes (Fig. 5.8).

5.4.5 Buckwheat Millet (Kuttu)

Buckwheat, also known as Kuttu in India, is a versatile and nutritious pseudo-cereal that is often considered a part of the millet family. In Indian cuisine, buckwheat (Kuttu) is commonly used during the Navratri fasting period and is often prepared into dishes such as puris, rotis, pancakes, and porridge (Fig. 5.9).

Here is some information about buckwheat millet and its potential health benefits:

Diabetic-Friendly and Blood Pressure Reduction: Buckwheat is low on the glycaemic index, which means it releases glucose into the bloodstream at a slower rate, making it a suitable choice for individuals with diabetes. It contains compounds that have been associated with blood pressure-lowering effects, potentially contributing to better cardiovascular health.

Cardiovascular Health and Weight Management: Buckwheat is rich in flavonoids, particularly rutin, which has antioxidant and anti-inflammatory properties. These compounds may help promote cardiovascular health and reduce the risk of

Fig. 5.9 Buckwheat Millet (KUTTU)

heart disease. The high fiber content in buckwheat can contribute to satiety and help with weight management.

Protection Against Breast Cancer, Asthma, and Gallstones: Some studies have suggested that certain compounds found in buckwheat may have anti-cancer properties and potentially protect against breast cancer. Buckwheat may also have a protective effect against asthma in children, although further research is needed to establish a definitive link. It has been suggested that the consumption of whole grains, including buckwheat, may lower the risk of gallstones.

5.4.6 Amaranth Millet (Rajgira/Ramdana/Chola)

Amaranth millet, also known as Rajgira, Ramdana, or Chola, is a nutritious type of millet that offers several health benefits. Here is some information about amaranth millet and its benefits:

Nutritional Value: Amaranth millet is rich in protein, making it a valuable source of plant-based protein. It is high in dietary fiber, which aids in digestion, promotes a healthy digestive system, and helps in maintaining a feeling of fullness. Amaranth millet contains essential minerals such as calcium, iron, magnesium, phosphorus, and potassium. It also provides vitamins, including vitamin A, vitamin C, vitamin E, and B vitamins.

Health Benefits: The protein content in amaranth millet helps in muscle development, repair, and overall growth. The dietary fiber present in amaranth millet supports healthy digestion and can aid in weight management. Amaranth millet contains antioxidants that may help reduce the risk of chronic diseases and provide anti-inflammatory benefits. It is believed to help lower cholesterol levels and reduce the risk of cardiovascular diseases. The presence of calcium and other minerals in amaranth millet supports bone health.

Hair and Skin Benefits: Amaranth millet is associated with potential benefits for hair health, as it contains nutrients that may help in fighting greying and hair

Fig. 5.10 Amaranth Millet (Rajgira/Ramdana/Chola)

loss. The presence of vitamins and minerals in amaranth millet can contribute to healthy skin and overall well-being.

Culinary Usage: Amaranth millet can be used in various culinary preparations, such as porridge, soups, salads, and baked goods. It can also be ground into flour and used to make bread, pancakes, and other gluten-free baked goods (Fig. 5.10).

5.4.7 Little Millet (Moraiyo/Kutki/Shavan/Sama)

Little millet, also known as Moraiyo, Kutki, Shavan, or Sama, is indeed one of the prominent millet varieties. Here is some information about little millet, including its nutritional value and culinary usage:

Nutritional Value: Little millet is rich in essential nutrients such as vitamin B, calcium, iron, zinc,

and potassium. It is a good source of dietary fiber, which aids in digestion and promotes a healthy digestive system. Little millet is gluten-free, making it suitable for individuals with gluten intolerance or celiac disease. It has a lower glycaemic index compared to rice, which means it causes a slower and steadier increase in blood sugar levels after consumption.

Culinary Usage: Little millet is widely used in traditional dishes, especially in the southern states of India, such as Tamil Nadu, Karnataka, Andhra Pradesh, and Kerala. It can be cooked and used as a rice substitute in various preparations, including pilafs, biryanis, porridge, and desserts. Little millet can also be ground into flour and used to make rotis, dosas, idlis, and other baked goods.

Health Benefits: Due to its nutritional composition and low glycaemic index, little millet is considered a healthier alternative to rice. It provides sustained energy and helps in weight management as it is rich in dietary fiber, which promotes a feeling of fullness. Little millet's nutrient profile supports bone health, blood circulation, immune function, and overall well-being.

Geographical Usage: Little millet is predominantly cultivated and consumed in India, particularly in the southern states. It is also grown in other countries such as Nepal, China, Myanmar, and some African regions (Fig. 5.11).

5.4.8 Kodo Millet (Barnyard Millet)

Barnyard Millet is Kodo millet, also known as barnyard millet, is a type of millet that has both nutritional and medicinal value. Here is some information about its medicinal value, area and production, and the geography of production:

Medicinal Value: Kodo millet is gluten-free, making it suitable for individuals with gluten intolerance or celiac disease. It is rich in dietary fiber, which aids in digestion and helps prevent constipation. Kodo millet is considered to have a low glycemic index, which means it releases glucose into the bloodstream at a slower rate and helps in managing blood sugar levels. It is a good source of essential nutrients, including B vitamins, iron, calcium, magnesium, and phosphorus. Kodo millet is also known to possess antioxidant properties and may have potential health benefits, such as reducing oxidative stress and inflammation.

Area and Production: Kodo millet is cultivated in various regions of India, particularly in states like Karnataka, Tamil Nadu, Andhra Pradesh, Telangana, Maharashtra, and Odisha. The production of kodo millet in India is estimated to be around 70,000 to 90,000 metric tonnes annually.

Geography of Production: Kodo millet is primarily grown in India but is also cultivated in other parts of the world. Besides India, kodo millet is grown in countries such as Nepal, China Myanmar, Bangladesh, Sri Lanka, and parts of Africa (Fig. 5.12).

5.4.9 Broomcorn Millet

Popularly known as Chena in India, Broomcorn helps balance blood sugar levels as it has a low glycemic index. It is a good option for diabetics to be incorporated into a daily diet. Switching to a diet with Millet can be a good transformation as far as nutrition is concerned. One can shop for Millets online and get a hand on the organic options available from different brands (Fig. 5.13).

Fig. 5.11 Little Millet (Moraiyo/Kutki/Shavan/Sama)

Fig. 5.12 Kodo Millet (Barnyard Millet)

Fig. 5.13 Broomcorn Millet

5.4.10 Kodo Millet

Area: The specific area under kodo millet cultivation in India is not readily available. However, it is cultivated in various states across the country, including Karnataka, Tamil Nadu, Andhra Pradesh, Telangana, Odisha, Maharashtra, and Madhya Pradesh. **Production:** The annual production of kodo millet in India is estimated to be around 70,000 to 90,000 metric tonnes.

WORLD:

Area: Kodo millet cultivation is not limited to India alone. It is also grown in other parts of the world, including countries in Africa, Asia, and some regions of South America. **Production:** The global production of kodo millet varies, but the specific figures for production in tonnes are not readily available on a global scale (Fig. 5.14).

Kodon Millet, is a digestible variant with higher amounts of lecithin amino acid. It has a significant effect on strengthening the nervous system. Kodo is a fantastic source of B vitamins, especially niacin, B6, and folic acid, among other vitamins and minerals. It contains calcium, iron, potassium, magnesium, and zinc minerals.

Fig. 5.14 Kodo Millet

Being a gluten-free millet, it is great for gluten-intolerant individuals. It can relieve cardiovascular disorders such as high blood pressure and cholesterol levels when eaten regularly by postmenopausal women.

References

Chellappan, B. V., & Peramaiyan, R. (2024). Unleashing the potential of millets: a comprehensive review of its nutritional, therapeutic, and genomic attributes. *Journal of Biological Research, 97*.

FAO. (n.d.). Retrieved from https://www.google.com/search?q=fao+statistics+2021+great+millets&tbm=isch&chips=q:fao+statistics+2021+great+millets,online_chips:pearl+millet:sdamxe8WqYw%3D&usg=AI4_kTAol_x62BZcZE6LzRzXEqIDfZigA&sa=X&ved=2ahUKEwiB8POJsLL9AhV51TgGHWeuD6IQgIoDKAN6BAgLEB0&biw=731&bih=343&dpr=2.63

Ganpat Louhar, R. S., Bana, V. K., & Kumar, H. (2020, December). Nutrient management technologies of millets for higher productivity and nutritional security. *The Indian Journal of Agricultural Sciences, 90*(12), 2243–2250. https://doi.org/10.56093/ijas.v90i12.110267

Grand View Research. (2030). *Millet Market Size, Share & Growth Analysis Report.*

International Year of Millets. (n.d.). Retrieved from https://www.google.com/search?q=international+year+of+millets+2023&oq=international+&aqs=chrome.2.0i67i433j69i57j0i67i131i433j0i67l2j69i60j69i61j69i65.9265j0j4&sourceid=chrome&ie=UTF-8

Mahajan, A. (2023). *The Magic of Millets: India's Super Grain Goes Global*. Retrieved from https://trendvisionz.com/lifestyle

Millet. (n.d.). Retrieved from https://en.wikipedia.org/wiki/Millet

MilletStats.com. (n.d.).

Production of Millets. (n.d.). Retrieved from https://www.google.com/search?q=area+production+and+productivity+of+millets+in+world&source=lnms&tbm=isch&sa=X&ved=2ahUKEwj6r9uvo7L9AhVz_3MBHcPZAfIQ_AUoAXoECAEQAw&biw=731&bih=343&dpr=2.63

ResearchGate. (n.d.). Retrieved from https://www.researchgate.net/figure/State-wise-area-production-and-yield-of-pearl-millet-in-India_tbl1_349313137

TATA AIG Team. (2024). *Health benefits of Ragi*. TATA AIG Team.

An Exploration on the Influence of Millet Consumption on Attaining Sustainable Development Goal 2

6

Vijay Kulkarni

Abstract

Despite outstanding progressions in the fields of farming and medication, the crucial reasons for destitution and unhealthiest keep on being troublesome issues. This might be on the grounds that we don't completely figure out how custom, culture, and the financial matters collaborate to influence various parts of wellbeing. Before, blended crop horticulture and the maintainable act of developing millets through regular and incorporated cultivating were fundamental to maintaining a decent lifestyle. Lamentably, this precious custom has been ignored after some time, to a limited extent on the grounds that to different variables and an absence of information. Regardless, this offers a great chance to revaluate millets' expected as far as their wellbeing benefits as well as their good consequences for the climate and horticultural strength. The Indian government (GoI) has executed various drives and projects pointed toward advancing millets. Moreover, aligning antiquated exercises with logical techniques and distributing about them won't just assist with investigating forward, yet in addition brace the connection between individuals, science, and their social legacy. This study examinations the progressions in the field, as well as the creation, efficiency, and utilization of millets in India at the public and state levels. It likewise takes a gander at whether the stockpile will be sufficient to satisfy the requests soon. The exploration likewise distinguishes SDG-2 and shows that in the forthcoming year 2030, there would be a net shortage on account of jowar and ragi yet an excess on account of bajra and little millets. It is feasible to increment creation in the close to term without extending the land by helping the dispersion of ensured seed, acquiring prevalence for new sorts, and improving harvest the executives methods.

V. Kulkarni (✉)
Director Corporate & Executive Program, Ajeenkya D Y Patil University, Pune, Maharashtra, India

C. B. Singh et al. (eds.), *Global Millets Production for a Sustainable Future*,
https://doi.org/10.1007/978-3-031-88157-2_6

Keywords

Millets · Sustainable Development Goal-2 · Production · Demand · Supply

6.1 Introduction

A group of little cultivated grains known as millets has been cultivated for millennia all through a large part of the world. They are a fabulous wellspring of supplements, plentiful in nutrients, minerals, and proteins, and high in fiber. In view of their san's gluten characteristics, they are ideally suited for individuals who have celiac illness or other aversion to gluten. To create bread, cakes, and pasta, millet can be squashed into flour or cooked entire as a porridge.

For centuries, in excess of 500 million individuals in Asia and Africa have depended on millets, which are developed in excess of 130 nations, as a staple food. Perhaps the earliest harvest to be trained in India was millets. Aside from their various wellbeing benefits, millets are harmless to the ecosystem because of their negligible water and information prerequisites. The Public authority of India (GoI) has given millets main concern as a result of their tremendous potential to help occupations, support rancher pay, and give food and nourishing security to individuals all over the place. Completely aim on making more interest and greater progression, millets were rebranded as "Nutri Grains" in April 2018 and the year 2018 was declared as the Public Year of Millets.

"Zero longing" is the target of Reasonable Improvement Objective 2. It is among the 17 Sensible Improvement Targets that the Gathered Nations set forth in 2015. "End hunger, achieve food security and further created sustenance and advance sensible agribusiness" is the authority communicating. To take care of the 800 million individuals in the world today, a critical upgrade of the food and rural frameworks overall is required. By focusing on millet creation, it very well may be attainable. Dryland makes up around 40% of the world's territory surface. The best harvest for dryland farming is millets.

These Nutri oats are yearly yields that are rainfed for a brief timeframe (75 to 120 days). They excel on low-richness, shallow soils that reach in pH from acidic to basic. It very well might be developed in very blistering environments with little precipitation and low water necessities. These require little upkeep, are impervious to most diseases and irritations, and can endure dry spells. Contrasted with C3 plants, these C4 plants have a superior photosynthetic proficiency and can change over CO_2 into starches. Millets are environment versatile yields and Nutri oats. It guarantees people's induction to food, food, and money related security. Millets are superfoods that are wealthy in gigantic degree and micronutrients. They are wealthy in bioactive parts and contain non-debilitating polysaccharides, sans gluten proteins, high dissolvability content, high cell fortresses, and a low glycemic record. B enhancements and beta-carotene are both plentiful in it.

The Latin word "Milum," which importance grain, is where "Millet" first appeared. The Ponceau family, furthermore implied as the grass family, consolidates the millet social event of oats. There are a couple of collections of millet, with assortments in assortment, surface, appearance, size of grain, and species. These are isolated into two classes considering grain size:

Large (Major) Millets: Jowar (Sorghum), Bajra (Pearl Millet), Finger Millet (Ragi). Proso Millet (Cheena) and Foxtail Millet (Kagni).

Small Millets: Kodo Millet (Kodra), Homestead Millet (Sama), Natural hued top Millet (Hari Kagni), and Little Millet (Kutki).

6.1.1 Background of the Study

Aside from the Northeastern regions of Himachal Pradesh and Jammu and Kashmir, most states in India—including Maharashtra, Karnataka, Andhra Pradesh, Madhya Pradesh, Gujarat, Rajasthan, Uttar Pradesh, and Tamil Nadu—create bajra and jowar. Both can be planted as Rabi (October–February) and Kharif (July–November) crops. In India, these yields are accessible in conventional assortments. Taking everything into account, they contrast enormously. They can be developed as blended, intercropped, or solitary yields. The length of the yield goes from 90 to 120 days. It is feasible to establish Jowar-Arhar and Jowar joined with different heartbeats, Bajra, and different grains. The dirt fruitfulness is expanded by pivoting crops by establishing mung and afterward jowar. One more method for developing bajra is as a blended yield.

One of Karnataka's most huge oats is finger millet, or ragi. It develops as summer and Rabi crops in Southern India yet for the most part as a Kharif crop in Northern India. It might flourish in soil that is pretty much as high as 11 pH and is antacid. The Ragi crop goes on for a sum of 135 days. It is developed as a slow time of year crop in rice neglected, as a blended harvest in with oilseed, bajra, beats, and Southern India's single crop. Both tropical and calm environments support the development of foxtail (Italian) millet. In Southern India, it develops all year. The yield goes on for eighty to one hundred distrained conditions are additionally used to develop Farm millet and Little millet. Both are versatile to waterlogging and drought. Browntop, Kodo, and Proso millets are particularly impervious to dry spell. Of the multitude of millets, Browntop has the briefest span—70–75 days.

Also, millets are created in flooded fields. For millets, one to two ploughings are adequate. For each kind of millet, there is an alternate speed of cultivating. It takes three to four downpours to become these crops. Dribbling or a seed drill are utilized for the planting system. Just insignificant measures of phosphatic or nitrogenous manures are required. Pesticide use is either expected at an insignificant or not the least bit. The stem and leaves are utilized as creature feed, and the actual panicles hold grains.

6.1.2 Grain Cereals and SDG 2: No Hunger

The worldwide goal of zero hunger is the focus of SDG 2. Millets have the potential to be quite helpful in accomplishing this goal since they are incredibly nutrient-dense and packed with important vitamins and minerals that are frequently absent from the diets of those who are impoverished.

Seeing this potential, the government of India, for instance, is promoting millets as a way to combat malnutrition.

Millets are a superb wellspring of protein and are effortlessly processed, making them an ideal nibble for small kids and infants. They are an economical dietary option for individuals in all financial gatherings since they are likewise a great wellspring of energy and can fulfil crave quite a while.

A review led by the Worldwide Yields Exploration Foundation for the Semi-Dry Jungles (ICRISAT) recommends that up to 24 million instances of hindering and up to 24 million instances of unhealthiest might be stayed away from assuming that millets were supported in Indian kids' eating regimens. These characteristics demonstrate the way that millets can actually add to the battle against hunger and the achievement of specific SDG 2 targets.

6.2 Research Objectives

- This study aims to examine the historical production patterns of millets in India from 1980 to 2020 by analysing changes in area, productivity, and production for several types of millets.
- The future supply of millets in India is projected using compound annual growth rates (CAGR), taking into consideration four distinct scenarios that are likely to have varying production trajectories.

6.3 Literature Review

Tanuja Manoj (2023) guaranteed that by laying out a general public in which nobody is abandoned, the SDGs give as a manual for accomplishing a superior and more feasible future for all. The 2030 Plan for Economical Turn of events, which revolves around the Reasonable Improvement Objectives, has a worldwide viewpoint and lays out the preparation for defending the desires and assumptions for coming ages. One of the targets, or Objective No. 2, is to "accomplish zero yearning." To end hunger, achieve food security, further foster sustenance, and advance affordable cultivating is the communicated objective of this mission.

Pandey et al. (2023) read up that vital for its capacity to definitively pinpoint the issues blocking the boundless utilization of millet, to give a numerical structure to determining the proposed short-and long-haul techniques for millet advancement, and to offer a guide for once again introducing millet to both metropolitan and rustic weight control plans. The review's decisions support a coordinated system to further develop esteem chain productivity and the making of an exhaustive structure for millet advancement that takes manageability in the space of economics, the climate, and wellbeing into account.

Patil et al. (2023) reintroduced millets into the Indian diet is fraught with difficulties, chief among them being the ubiquitous effect of commercials that currently promote cereal-based items, domestic kitchen politics, market-driven agriculture methods, and economic considerations. Despite these challenges, the goal of the current publication is to offer insights into creating a roadmap that makes it possible to integrate millets into sustainable development goals.

Subhash et al. (2022) treated the soil farming waste can act as a substitute for compound manures that require a ton of energy. In addition, esteem added rural waste can possibly be utilized as a feedstock for modern and animals products. Notwithstanding, there isn't sufficient careful data about Agri-squander the executives in the writing. Accordingly, the ongoing review analyzed the latest advancements in successful Agri-squander the board innovation. The latest examination will help strategy creators and specialists in growing earth sound buildup the executives techniques to empower the rural delivering area change to a green economy.

Udenigwe (2021) examined the data that is currently available on the relationship between malnutrition and food processing techniques, food matrix components, and nutrient bioavailability. Crops that are underutilised or neglected were also mentioned as being crucial to reducing the triple load.

Satyavathi et al. (2021) guaranteed that since pearl millet is an environment versatile yield, it can help cultivating networks in dry areas by diminishing the adverse consequences of environmental change and by possibly expanding their pay and food security. Because of its solid root structure, pearl millet can endure different ecological burdens, including water shortage. It is less subject to fake manures and shows extraordinary photosynthetic effectiveness, great creation, and development in low-supplement soil conditions. In light of these characteristics, pearl millet is a harvest that is liked for development in dry and semi-dry locales of the world. In any case, contrasted with other principal oats, less exploration has been finished on the environment versatile attributes of pearl millet. India's legislative and confidential areas have fostered various mixtures and assortments of pearl millet throughout the course of recent years.

FAO Distribution (2016) expressed that—Among the commendations of heartbeats are their immense geological reach, high healthy benefit and low water prerequisites, their special capacity to self-treat, (adding fundamental nitrogen to cultivate land and further developing yields en route), alongside keeping up with their medical advantages over a long timeframe of realistic usability. Beats are a steady weapon against yearning and unhealthiest wherever for these reasons. "Heartbeats are a certified superfood for the future," it continues. It's undeniably true that heartbeats would without a doubt fulfil the twin prerequisites of sustenance and supportability to battle the high weight of undernutrition.

Hariprasanna et al. (2014) stressed the meaning of millets in getting dietary security "The groundwork of rainfed cultivating, which supports 60% of India's cows populace and gives a living to almost half of the country's provincial labour force, is the development of millet." They included 12 that, Significantly, millet is a versatile harvest that can endure dry season and is profoundly versatile to a scope of agroclimatic challenges. Because of their more noteworthy versatility, they are in many cases utilized in uneven and semi-parched tropical areas for minor horticulture. Since millets are C4 plants, their photosynthetic framework is definitely more proficient than that of C3 plants, like wheat and rice.

6.4 Research Methodology

The examination utilized information on crop yield accumulated from the Directorate of Financial matters and Insights between 1970–1980 and 2010–2020, as well as different rounds of family utilization spending information from the Public authority of India's Public Example Review Office (NSSO) between 1987–1988 and 2011–2012. Compound yearly development rates were utilized to figure the millets' future stock (CAGR). To get a scope of future inventory, four situations have likewise been thought about for projection.

The behaviouristic methodology, which depends on populace development and changes in utilization designs because of changes in per capita pay in a developing economy, estimated regarding utilization use flexibilities, was utilized to project the immediate interest (family interest) for millets. We utilized the spending flexibilities for coarse cereals that Kumar et al. processed. The gauge interest from the NSSO 68th cycle (2011–2012) was utilized to anticipate interest for both provincial and metropolitan families. We have at long last shown up at the immediate public interest for various millets by collecting the interest of families in both rustic and metropolitan regions. Moreover, the roundabout interest for coarse grains was assessed to represent 32.47% of the general stock, considering interest for seed, feed, squander, and modern use. Ultimately, the general interest for not entirely settled by adding the immediate and roundabout interest.

6.5 Result and Discussion

6.5.1 Production Trends

Like Table 6.1, the region of all millets viable diminished consistently somewhere in the range of 1980 and 2020. Because of a 2.5-overlap improvement in efficiency, the region planted with pearl millet has diminished by roughly 34% while creation has expanded by roughly 63%.

Jowar, which had a huge region portion of 16.09 mha during the 1970s, declined to 6.02 mha, a diminishing of almost 64%. This prompted a 44% decline underway notwithstanding a 51% gain in efficiency. The efficiency has expanded by practically 68%, while the ragi region has diminished by almost half, bringing about a result reduction of almost 23%. Small millets saw the greatest region decrease, of up to 85%, which prompted a 77%decrease underway despite the fact that the yield

Table 6.1 India's millets' decadal averages for area, productivity, and production

	Area	Production Pearl millet	Productivity	Area	Production Jowar	Productivity	Area	Production Ragi	Productivity	Area	Production Small millets	Productivity
1980–1990	11.02	5.30	0.50	16.09	11.1	0.70	2.47	2.60	1.08	3.40	1.38	0.6
1990–2000	10.08	6.75	6.69	12.25	10.20	0.80	1.90	2.50	1.5	1.90	0.90	0.50
2000–2010	9.35	8.07	0.88	9.03	7.45	0.86	1.50	2.10	1.40	1.15	0.50	0.49
2010–2020	7.90	9.2	1.17	6.02	5.5	0.90	1.17	1.80	1.60	0.65	0.40	0.65

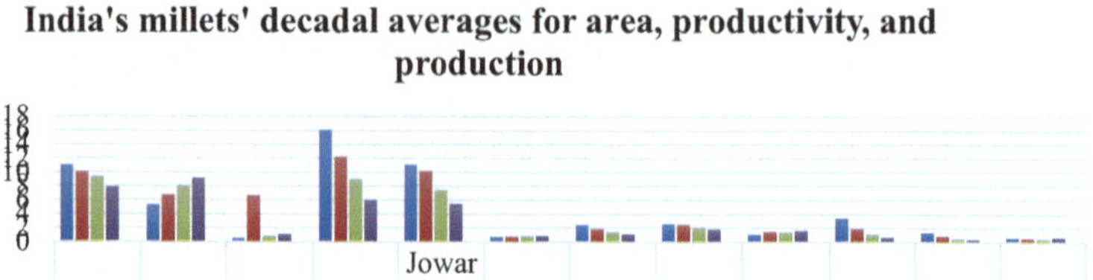

	Area	Production Pearl Millet	Productivity	Area	Production	Productivity	Area	Production Ragi	Productivity	Area	Production Small Millets	Productivity
1980-1990	11.02	5.3	0.5	16.09	11.1	0.7	2.47	2.6	1.08	3.4	1.38	0.6
1990-2000	10.08	6.75	6.69	12.25	10.2	0.8	1.9	2.5	1.5	1.9	0.9	0.5
2000-2010	9.35	8.07	0.88	9.03	7.45	0.86	1.5	2.1	1.4	1.15	0.5	0.49
2010-2020	7.9	9.2	1.17	6.02	5.5	0.9	1.17	1.8	1.6	0.65	0.4	0.65

Fig. 6.1 India's millets' decadal averages for area, productivity, and production

expanded by 58% during a similar time span. Subsequently, one might say that the expansion in millets' efficiency has not balanced the diminishing in that frame of mind, except for pearl millet.

As Table 6.2 represents, millet creation in India is for the most part focused in a small number of locales across a small number of states. A large portion of the pearl millet yield's region (around 70%) and creation (around 62%) are shared by Rajasthan and Uttar Pradesh. Then again, Gujarat stands first in a shockingly lengthy timespan of viability (2.31 t/ha) trailed by Tamil Nadu (2.30 t/ha) and Madhya Pradesh (2.23 t/ha). High-conveying states like Uttar Pradesh and Rajasthan were found to have yields of 1.97 t/ha and 0.92 t/ha, autonomously. For TE 2019, the ordinary public still hanging out there to be 1.24 t/ha. Maharashtra, Karnataka, and Rajasthan are the states with the largest acreage and output of jowar, however they do not have the highest productivity levels. Karnataka is the top state in terms of ragi production and area, accounting for almost 60% of the total. It also has the second-best productivity (1.45 t/ha), but slightly behind the 1.45 t/ha national average. When it comes to the area and output of tiny millets, Madhya Pradesh is the state leader. For tiny millets, the area under cultivation is more broadly dispersed, whereas for ragi, it is heavily concentrated. With very few exceptions, the area and productivity are decreasing in the top states. However, given that these states' productivity levels are relatively low, there is room for improvement. In addition, because millet is grown in such a small number of states, it is simpler to plan and focus programmes specifically for improving millet output.

The region shares, creation share, efficiency (TE 2018–2020), and yearly development rate somewhere in the range of 1980 and 2000 are displayed in the figures in enclosure. The development rate was figured between 2013–2014 and 2018–2020.

Table 6.2 Top millet delivering states

Millets	Top three states		
	Area	Production	Productivity
Bajra	Rajasthan (57.03; −0.08)	Rajasthan (42.43; 3.25)	Gujarat (2.31; 2.96)
	Uttar Pradesh (12.30; −0.34)	Uttar Pradesh (19.22; 2.35)	Tamil Nadu (2.30; 2.31)
	Maharashtra (10.12; −1.50)	Gujarat (10.13; −0.29)	Madhya Pradesh (2.23; 2.82)
Jowar	Maharashtra (46.9; −1.62)	Maharashtra (36.17; −0.74)	Madhya Pradesh (2.03;1.95)
	Karnataka (20.20; −1.62)	Karnataka (22.45; −0.73)	Andhra Pradesh (1.87; 3.03)
	Rajasthan (11.24; −1.27)	Rajasthan (8.74; −0.15)	Gujarat (1.33; 3.10)
Ragi	Karnataka (61.35; −0.83)	Karnataka (61.22; 0.80)	Tamil Nadu (2.93;1.83)
	Uttarakhand (9.71; −2.45[a])	Tamil Nadu (15.03; −1.25)	Karnataka (1.44; 1.61)
	Maharashtra (8.57; −1.71)	Uttarakhand (8.90; −1.61[a])	Uttarakhand (1.33; 0.85[a])
Small millets	Madhya Pradesh (26.2; −5.04)	Madhya Pradesh (26.01; −3.40)	Uttarakhand (1.33; 0.65[a])
	Chhattisgarh (16.34; −6.95[a])	Uttarakhand (19.05; −0.80[a])	Tamil Nadu (1.22; 1.01)
	Uttarakhand (10.93; −3.64[a])	Tamil Nadu (7.23; −5.59)	Gujarat (1d.13; 0.70)

[a]Uttarakhand and Chhattisgarh were part of Uttar Pradesh and Madhya Pradesh respectively before 2000

6.5.2 Supply and Demand for Millets

Four circumstances have been contemplated for the stock assumption for millets. Under situations 1, 2, and 3, it was speculated that creation would continue to increment according to improvement rates over the beyond 5, 10, and 15 years, independently. In situation 4, it was acknowledged that outcome would stretch out according to locale consistent and past yield advancement plans. To put it immediately, situations 1–3 idea about the area's reduction, while circumstance 4 (the confident situation) acknowledged that there wouldn't be any more locale decline. Under conditions 1–3, the inventory of bajra, jowar, ragi, and little millets by 2029 may be 9.43–9.61, 2.47–2.97, 1.1–1.40, and 0.35–0.41 Mt, autonomously. Then again, under a certain situation, the inventory of bajra, jowar, ragi, and little millets could expansion to 9.91, 4.15, 1.32, and 0.67 Mt, freely.

Three scenarios were analysed for the family interest (direct) projection. The genuine interest gauges for situation 1 was determined by considering the pattern interest for various kinds of millets in 2011. It is accepted that the accessibility of prepared to-eat esteem added merchandise and developing public consciousness of millets' wellbeing benefits are the principal explanations for the ongoing flood sought after for millets for homegrown use. Along these lines, two additional scenarios were analysed, accepting expansions sought after over projection utilizing benchmark interest of 5% (scenario 2) and 10% (scenario 3). By 2029, family demand should be 2.67, 2.57, 1.09, and 0.12 million tons for bajra, jowar, ragi, and

Table 6.3 Millets' net surplus or loss in different scenarios (Mt)

Year	Bajra	Jowar	Ragi	Small millets
2023	3.22 to 3.52	−0.44 to −0.15	−0.15 to −0.02	0.17 to 0.19
2025	3.30 to 3.63	−0.70 to −0.41	−0.20 to −0.06	0.14 to 0.19
2027	3.40 to 3.70	−0.92 to −0.62	−0.27 to −0.13	0.16 to 0.17
2029	3.50 to 3.80	−1.13 to −0.81	−0.30 to −0.16	0.12 to 0.20

minor millets, separately. The total amount of bajra, jowar, ragi, and little millets mentioned by 2029 would be 2.9–3.1, 2.8–2.9, 1.15–1.9, and 0.2 Mt, independently, if there is a 5-to 10-percent rise famous. To decide the whole interest and decide the interest supply hole, an expected measure of backhanded request (interest for seed, feed, squander, and modern purposes) was additionally thought of. By ascertaining the distinction, the scope of the net not entirely set in stone between possible pairings of organic market in the previously mentioned conditions. Table 6.3 shows that by 2029, there may be a net excess of 3.50 to 3.80 Mt. for bajra and 0.12 to 0.20 Mt. for little millets. However, there might be a net shortfall for jowar and ragi in the range of −1.13 to −0.81 and − 0.30 to −0.16 Mt., respectively; as a result, steps might be made to guarantee a steady supply.

6.6 Conclusion and Recommendation

To ensure food security, improve nourishment, and protect ranchers' vocations while safeguarding the climate, SDG-2 looks to progress supportable horticulture rehearses (SAPs). Expanding agrarian efficiency, helping smallholder ranchers, and progressing versatile and supportable cultivating frameworks are the fundamental goals. Furthermore, it jellies hereditary variety, improves the strength of the land and soil, and helps in the decrease of the use of hazardous synthetic compounds a pesticide. This study thoroughly examines the current state of millet production and consumption as well as associated issues that can help researchers and policymakers formulate plans to bring millets back to their former glory. For the benefit of people and the environment, both production and consumption ought to increase. Raising consumer knowledge of the nutritional and environmental benefits of millet and creating products that meet their needs—including prepared foods—will encourage usage. There will shortly be a shortage of jowar and ragi due to India's present production and consumption patterns. Increasing production will help close this gap. Since the majority of millet production occurs in a small number of states with low productivity, focusing on increasing productivity in these states can lead to an immediate boost in production. Millets are mostly cultivated as poorly managed rainfed crops. Therefore, increasing the use and cultivation of high-yielding cultivars, providing irrigation at least at the crop's crucial stage, and implementing sound cropping techniques will boost output and productivity.

6.6.1 Recommendations

The accompanying ideas are made considering the review's outcomes about the job of millets in guaranteeing nourishment security in India:

- Agricultural extension services, new seed types, and better farming practices can all contribute to promoting pearl millet production, which is worth doing because it increases productivity significantly with less acreage needed.
- Incentives for farmers, research into high-yielding cultivars, and marketing tactics to raise demand should be implemented to revitalise the cultivation of jowar and ragi, as both their acreage and output have declined.
- Cultivating a Variety of Millets: Promoting the growing of minor millets can help diversify crops and make agricultural systems more resilient. Awareness campaigns and specialised support programmes could accomplish this.
- Efficiency Upgrade: Millet creation can be made more interesting to ranchers by putting resources into innovative work, water system foundation, and soil well-being the board, all of which can increment absolute efficiency.
- The nutritious value of millets and their contribution to combating malnutrition should be the focus of more public education initiatives. Domestic demand for millets can be increased and a more stable market can be established through this.

References

Babu, S., et al. (2022). Exploring agricultural waste biomass for energy, food and feed production and pollution mitigation: A review. *Bioresource Technology, 360*, 127566.

FAO. (2016). *Pulses—nutritious food for a sustainable agriculture: 2016 International Year of Pulses*. FAO.

Hariprasanna, K., Gomashe, S., Ganapathy, K. N., & Patil, J. V. (2014). Millets for ensuring nutritional security. *Popular Kheti, 2*, 170–175.

Mensi, A., & Udenigwe, C. C. (2021). Emerging and practical food innovations for achieving the Sustainable Development Goals (SDG) target 2.2. *Trends in Food Science and Technology, 111*, 783–789.

Nesari, T. M. (2023). Celebrating International Year of Millets: Way towards holistic well-being. *Journal of Ayurveda Case Reports, 6*, 1.

Pandey, A., & Bolia, N. B. (2023). Millet value chain revolution for sustainability: A proposal for India. *Socio-Economic Planning Sciences, 87*, 101592.

Patil, P. B., et al. (2023). Millets: Empowering the society with nutrient-rich superfoods to achieve sustainable development goals. *Journal of Drug Research in Ayurvedic Sciences, 8*(Supply 1), S100–S114.

Satyavathi, C. T., et al. (2021). Pearl millet: a climate-resilient nutricereal for mitigating hidden hunger and provide nutritional security. *Frontiers in Plant Science, 12*, 659938.

7 Little Millet (*Panicum sumatrense*); A Sustainable, Delectable, and Healthy Alternative to Cereals Like Wheat and Rice

Chaudhary Mohd Kashim and Noopur Gautam

Abstract

Millet is an umbrella term for small grains that can be termed nutri-cereals and utilized for food, feed, and fodder. Little millet (*Panicum Sumatrense)* is a traditional grain in India that is grown in the semiarid and arid regions that are prone to drought. Little millet is grain-bearing crop belonging to the family Poaceae and the genus *Panicum*. With a low glycemic index of 50–65 and being gluten-free, *Panicum Sumatrense*, one of the minor underutilized grains, is excellent for people of all ages. It is mostly consumed for food, especially by people of economically weaker section. The grain is a relative of proso millet, although its seeds are significantly smaller than proso millet's. *Panicum Sumatrense* has a nutritional profile that is comparable to that of other cereals like wheat and rice. The millet is abundant in B vitamins and minerals like calcium, iron, zinc, and potassium. It feeds the body with necessary fats, the type that aids in weight loss. Polyphenols, flavonoids, and phenolic compounds are among the anti-oxidants found in little millet. *Panicum Sumatrense* contains antinutritional substances such as phytic acid, tannins, free phenolics, and protease inhibitors, which can be reduced by soaking, sprouting, or lactic acid fermentation. Nutraceuticals such phenolics, gama-amino butyric acid (GABA), lignans, resistant starch, sterols and phytates are also abundant in it. *Panicum Sumatrense* assists in the treatment of cancer, cataracts, diabetes, gastrointestinal problems, and other diseases. Dosa, idli, pongal, and kichadi are few dishes that can be made with little millet. This chapter emphasizes on little millet, a minor millet that shows great promise for food security, processing, and health benefits.

Keywords

Little millet · Underutilized · Nutraceuticals · Health benefits · Food security

C. M. Kashim (✉) · N. Gautam
Institute of Food Technology, Bundelkhand University, Jhansi, Uttar Pradesh, India

C. B. Singh et al. (eds.), *Global Millets Production for a Sustainable Future*,
https://doi.org/10.1007/978-3-031-88157-2_7

7.1 Introduction

Millets are a major food source in arid and semi-arid parts of the world. Millets are good sources of energy. They provide protein, fatty acids, minerals, vitamins, dietary fibre and polyphenols. Millets are one of the cereals asides the major wheat, rice, and maize. Millets are major food sources for millions of people, especially those who live in hot, dry areas of the world. They are grown mostly in marginal areas under agricultural conditions in which major cereals fail to give substantial yields (Amadou et al., 2013). The four major types are pearl millet (*Pennisetum glaucum)* which compromises 40% of the world production, foxtail millet (*Setaria italica*), Yang *et al.,* proso millet or white millet (*Panicum miliaceum*), and finger millet (*Eleusine coracana).* Minor millets include barnyard millet (*Echinochloa spp.*), kodo millet (*Paspalum scrobiculatum*), little millet (*Panicum sumatrense*), browntop millet (*Urochloa ramosa* = *Brachiaria ramosa* = *Panicum ramosum*), sorghum (*Sorghum* spp.) (ICRISAT, 2007; FAO, 2009; Adekunle, 2012) (Fig. 7.1 shows different types of millets).

Little millet (*Panicum sumatrense, Panicum miliare*) one of the minor millets named as kutki in hindi, sava in marathi, gajro in gujarati, sama in kannada, samai in tamil, and samalu in telugu. People of all ages can use it. It is a traditional crop

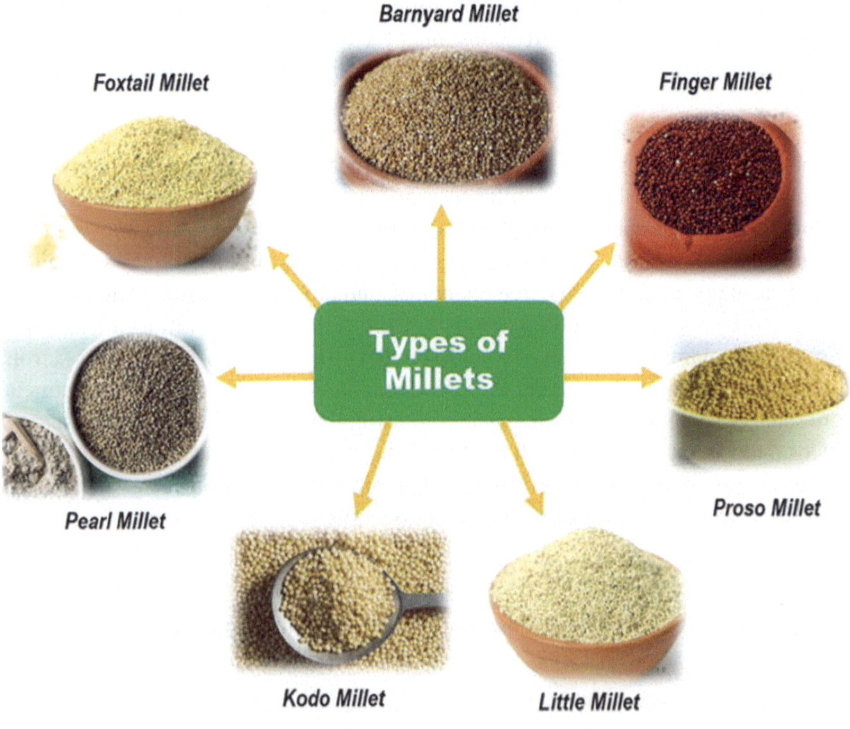

Fig. 7.1 Types of millets. (Source: Tiwari et al., 2022)

that is grown all across India. Although little millet and proso millet are related, little millet seeds are substantially smaller. It is a self-pollinating crop, Small, yellow and gluten free with a low glycemic index of 50–65 which is less than cereals like wheat and rice (Agrawal et al., 2023). Little millet is another reliable catch crop as it is resistant against extreme agro-climatic conditions. It has received comparatively little attention from plant breeders. The plant varies in height between 30 and 90 cm. It is mostly grown along with other millets, pulses and oilseeds. The seeds of little millet are smaller than those of common millet. Eating little millet in summer is the best way to beat the heat. Hence, it is called "cool food" (Swamy, 2023). Its easy digestible property makes it an ideal crunchy breakfast snack for all age groups (The Hindu, 2015) (Fig. 7.2 name of little millet in local languages used in India).

Little millet (*Panicum sumatrense* Roth. ex. Roem. and Schultz.) is one of the important small millets indigenous to Indian subcontinent (Nirmalakumari et al., 2010). It is fibrous millet next to barnyard millet. Little millet is highly nutritious

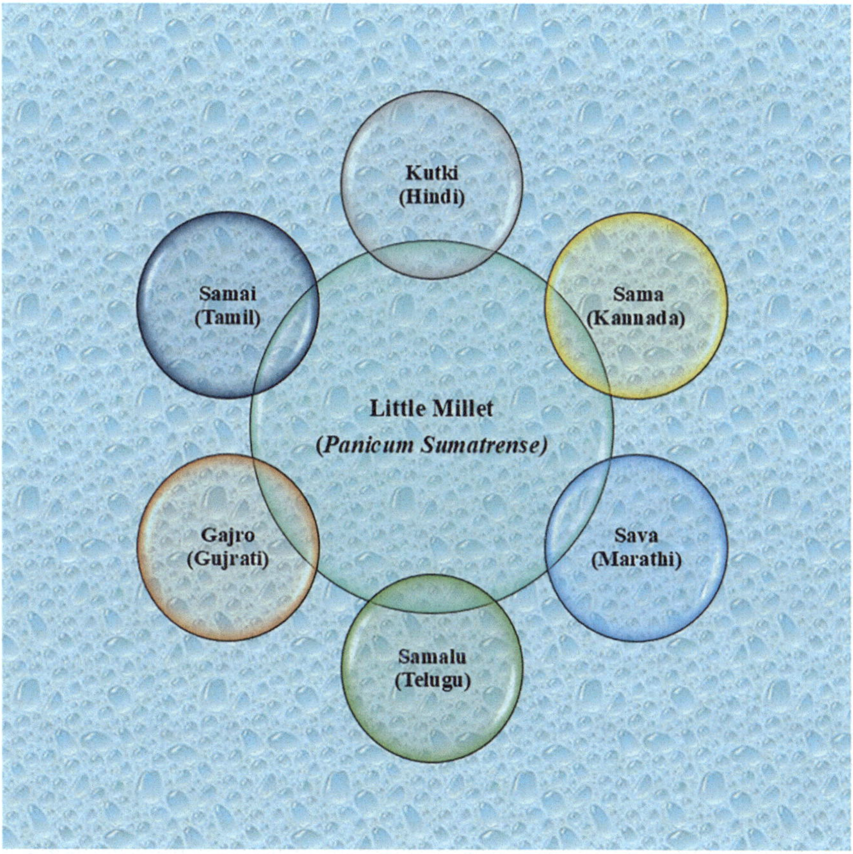

Fig. 7.2 Little millet name in local languages

and may be called little but it not less in its nutritional content. It has good source of B-vitamin, minerals like calcium, iron, zinc, potassium among others. It also provides essential fats to the body, the kind that helps in weight loss. Its high fiber content is yet another positive making it an ideal part of pongal or even kheer instead of rice (Reddy, 2017). Little millet belongs to the family Poaceae, sub-family Panicoideae, tribe Paniceae, Genus *Panicum* Species *Panicum sumatrense* (Saloni et al., 2018; Ladumore et al., 2021). Little millet (*Panicum sumatrense* Roth. Ex. Roemer and Schultes) is grown in India under various agro ecological situations. The time to maturity for most cultivars is about 90 days (Goron & Raizada, 2015). Little millet is grown throughout India up to altitudes of 2100 m, but it has only a little importance in different places (Indirani & Devasena, 2021). Little millet is a tetraploid with $2n = 4x = 36$ (Goron & Raizada, 2015). Little millet matures quickly and withstands both drought and water logging. Less genetic diversity occurs in the world collections of this species than appears among the other species and the grains are similar to that of rice. Perhaps very little of this species is grown outside of India (Bhat et al., 2018). Little millet is commonly available across the country as a whole grain.

7.2 Origin and Domestication

Millets are among the oldest crops domesticated and cultivated in the world for human food and animal fodder and their cultivation dates back to 8700–10,300 years ago. Various millet species were initially domesticated in different parts of the world like South Asia, East Asia, East Africa, and West Africa. However, they spread well beyond their initial area of domestication. Whereas, the earliest record of millets domestication and cultivation is of Foxtail millet and Proso millet from China around 3000–2000 BC and Indian valley of Kashmir is regarded as a place of integrated networks, where major trade of millets took place between Asia, Europe, and Africa (Kumar et al., 2021).

Archaeo-botanical studies of history of millet cultivation in India reported from the pre-Harappan culture at Rohira, the Late Harappan culture at Hulas in western Uttar Pradesh, in the same context at several sites in Guarajat (Rangpur, Surkodta, Rojdi) and in the Neolithic levels at Hallur (Karnataka), besides Pirak in Pakistan. These grains were grown in the Chalcolitihic culture at Ahar (south-eastern Rajasthan) and at Paunar in northern Maharashtra. "The origin and dispersal of millet cultivation in India" the evidence for the cultivation of jowar millet (Sorghum vulgare L.) at Rohira has been obtained from Period IA (pre-Harappan, ca. 2300–2000 B.C.) along with that of barley, wheat, lentil and horse gram (IAR, 1983–84: 188) and likewise (Gautam et al., 2023) (Fig. 7.3 highlights the close-up images of little millet panicles).

Between the Indo-Gangetic Plain to the north and the Vindhya Range to the south, the Bundelkhand region of the Indian state of Uttar Pradesh is hot and semi-humid. In Uttar Pradesh, it is made up of the following 7 districts: Banda, Chitrakoot, Hamirpur, Jalaun, Jhansi, Lalitpur, and Mahoba. The rainfall in this region is 50 to

Fig. 7.3 Panicles of little millet. (Source: https://www.crops.org/news/science-news/big-potential-little-millet/)

100 mm and the soil is partly black cotton and partly lateritic. In 1965 the district gazetteer of Jhansi said that in Bundelkhand, Lalitpur was particularly more dependent on small millets than other districts of the region (at that time Lalitpur was a subdivision of Jhansi). These grains generally included mandwa or ragi, kodon (*Paspanum scrobiculatum*), sawan (common millet), kakun (Italian millet) and kutki (little millet) and have always covered a considerable area in kharif. They were mostly grown in the Lalitpur subdivision partly because of inferior red soil which was suited to the cultivation of such crops and partly because of the low standard of cultivation. It is considered *Fikar* and *Samai* (Little Millet) are the traditional small millets of Bundelkhand, especially of Lalitpur district (Gautam et al., 2023).

Panicum sumatrense is grain-bearing crop belonging to the family Poaceae and the genus *Panicum* (Wikipedia). Little millet is one of the oldest crops domesticated in India and is well adapted to varied soil and environmental conditions. Although little millet, like all millet, is nutritionally superior to cereals, its use is limited. Hence, there is a need to restore the lost interest in little millet due to its potential nutritional qualities and health benefits. Little millet, *Panicum sumatrense* Roth. ex Roem. & Schult. (Syn. *Panicum miliare*) was domesticated in India from 5000 years (Johnson et al., 2019). Eastern Ghats of India are known to be the place of domestication of little millet as early as 2000 years ago and the crop is majorly cultivated in peninsular Indian states like Andhra Pradesh, Karnataka, Tamil Nadu, and Kerala. The crop is adapted to both dry and humid conditions and can be cultivated in drought-prone areas as well as water-logged conditions, as the crop matures early and withstands adverse conditions. The genetic diversity of this crop is very little because of its restricted cultivation in India, Sri Lanka, Nepal, and Myanmar, with India accounting for more than 98% of the area and production of little millet (Kumar et al., 2021). In India, the area under little millet cultivation has reduced from about 0.5 to 0.3–0.4 million hectares during 2001–2006 with a production of around 0.1 million tonnes (Ganapathy et al., 2017). Little millet is a quick growing crop possessing excellent rejuvenating capacity and is staple food for the low-income groups in some parts of world (Hiremath et al., 1990). As this crop is eco-friendly, it is suitable for fragile and vulnerable Agro-ecosystems growing in tropical and sub-tropical climates, elevated areas up to 2100 m along with plains and hilly regions (Kumari et al., 2010). Indian little millet has a short crop cycle, is mostly cultivated under natural rainfed irrigation and suitable for both dry and waterlogged conditions. It also grows well in adverse environments such as salty and dry soils as well as at high temperatures (Sivakumar et al., 2006). This crop gives consistent yields on marginal lands in drought prone arid and semiarid regions providing regional food stability. In Madhya Pradesh, kodo (70%) and little (24%), together, contribute to 94% of the total area under small millets. In Madhya Pradesh, there are two districts that have the highest area under the state: Dindori and Mandla for the production of kodo and Chhindwara and Dindori for the production of little millets.

Little millet is also a yearly crop having 30–90-cm hollow jointed stem with slender and robust base. The leaves are linear and can be 15–50 cm in length to 12–25 cm in breadth, while the nodes are glabrous. The seeds are grown during the period from June to July or between February and March. Its grain size is approximately 2.5 × 1.5 mm with 1.9 g of approximate kernel weight oval in shape, and the colour is creamy with shiny appearance. "Cool food" is another name for little millet because of the cooling impact it has on the body when consumed in summer time (Dey et al., 2022) (Fig. 7.4 showing the little millet grains).

Fig. 7.4 little Millet grains. (Source: https://nrootsfood.com/product/little-millet-grains/)

7.3 Botanical Description

Miliare is another name for little millet (Panicum sumatrense Roth. ex Roem. & Schult.). It is extensively grown as a cereal in western Myanmar, Nepal, and India. It is especially significant in India's eastern ghats, where it plays a significant role in tribal agriculture. Following a comparative morphological analysis of accessions, two races were identified. The size of race nanaplants ranges from 60 to 170 cm, with decumbent culms that root at lower nodes and then grow upright to produce blooming culms. The terminal inflorescences are erect, open, and 14–50 cm long. They have robust branches, which occasionally group together as they reach maturity. Plants belonging to the race robusta are erect or develop blooming culms from a short geniculate base. Flowering culms are 120–190 cm tall and robust. Terminal inflorescences are heavily branching, compact or open, and measure 20–46 cm in length. When mature, compact inflorescences become bent, while open inflorescences are virtually erect. The crop is mostly self-pollinated, with up to 3.5% of spontaneous cross-pollination occurring. The dehusked grain can be ground into flour or cooked and eaten like rice. The crop matures in two to five months and grows in environments that cannot support any other edible plant. This 5–2.5 m tall plant has sturdy stems that are somewhat branched and covered in long hair. Its hue is generally reddish or purplish. Stipules are absent, the petiole can grow up to 8 cm long, the blade is broadly ovate to rhomboid-ovate, and the inflorescence is large (up to 1.5 m) and complex, with many agglomerated cymes, an axillary spike, and terminal spikes that are pendant to erect; the bracts are 2–4 mm long, membranous, pale, and have a long awn. Sessile, unisexual blooms with five mucronate tepals and around one mm-long stamens; three stigmas top the female flowers. Fruit an ovoid long, circumscissile, almost smooth or slightly furrowed, abruptly narrowed to a short thick beak, 1-seeded (Saloni et al., 2018).

Little millets known as samai or kutki are short-duration millets and withstand both drought and water logging. A yearly crop, little millet has a hollow, jointed stem that ranges in length from 30 to 90 cm and a slender, robust base. The nodes are glabrous, and the leaves are linear, ranging in length from 15 to 50 cm to 12 to 25 cm in width. The seeds are planted either in February or in March, or in June and July. Its oval-shaped kernels weigh roughly 1.9 g and have a grain size of approximately 2.5 × 1.5 mm. Their hue is creamy and lustrous. Little millet is also known as "cool food" due to the cooling effect it exerts on the body during the summer (Dey et al., 2022). Little millet has a similar habit to proso millet, except that it is smaller. This annual herbaceous plant can reach a height of 30 cm (12 in) to 1 m (39 in.) with either straight or folded blades. The leaves are linear, with sometimes hairy laminae and membranous hairy ligules. The panicles range in length from 4 to 15 cm (1.6 to 5.9 in.), with an awn of 2 to 3.5 mm (0.079 to 0.138 in.). The grain is round and smooth, measuring 1.8 to 1.9 mm (0.071 to 0.075 in.) long. The seed measures 2–3 mm in length. Grain shapes range from elliptical to oval, with colours ranging from grey to straw-white (Wikipedia, 2023) (Fig. 7.5 illustrates the botanical description of little millet grain from seedling to grain).

Fig. 7.5 Botanical description of little millet. (Source: Swamy, 2023)

7.4 Agronomic Characteristics of Little Millet

Little millet (Panicum sumatrense) is a cereal crop with small seeds that is drought tolerant and may grow in marginal soils. It grows quickly, taking only 70–110 days to reach a height of 30–90 cm. *Panicum Sumatrense* is tolerant to a variety of agro-climatic conditions and may be grown in both rainfed and irrigated areas. Agronomic features that are noteworthy include: resistance to pests and diseases, low input requirements, adaptation to poor soil conditions, tolerance to drought, and short growth period.

7.5 Nutritional Profiling of Little Millet

Panicum Sumatrense, one of the minor underutilized grains, is ideal for individuals of all ages because it has a low glycemic index of 64 and is gluten-free (Anitha et al., 2021). As per Gopalan et al. (2010), little millet is a cheap source of energy

(341 kcal/100 g), protein (7.7 g/100 g), crude fiber (7.6 g/100 g) and minerals (1.5 g/100 g), particularly rich in iron (9.30 mg/100 g) compared to raw rice (0.7 mg/100 mg). Little millet has a respectable amount of additional nutrients like niacin, pantothenic acid, folic acid, calcium, iron, potassium, phosphorus, magnesium, and zinc. With food applications as nutraceutical and functional food ingredients for health promotion and illness risk reduction, little millet is a natural source of antioxidants as gallic acid, p-hydroxybenzoic acid, vanillic acid, caffeic acid, chlorogen acid, ferulic acid, and p-coumaric acid. It also contains γ-amino butyric acid (GABA), lignans, resistant starch, sterols and phytates exhibiting hypoglycaemic, hypercholesteraemic and hypolipidemic properties (Guha et al., 2015; Neeharika et al., 2020).

Kushwaha et al. (2018) notified that substantial retention of phytates has a beneficial effect on health due to the antidiabetic and anticancerous properties and acts as antioxidants. It helps to prevent constipation and heals most of the stomach problems. Its significant high fiber content aids in the reduction of fat deposits (Kumar et al., 2016) in the body and is also responsible for the movement of food and waste through the digestive system (Obafaye et al., 2018). Little millet is rich in cholesterol, when consumed increases good cholesterol in the body, suitable for growing kids and strengthens the body. Its complex carbohydrate digests slowly which is very helpful for diabetic patients (Gayatri 2015). Little millet and kodo millet have 37–38% of dietary fiber and regarded as nutraceutical crops (Hegde & Chandra, 2005). Fiber helps maintain sound digestion and satiates hunger quickly. Like barnyard millet and foxtail millet, little millet is rich in iron. The flavonoids present in the little millet act as antioxidants and plays many roles in the body's immune defence system. Although amino acids are poor in composition but little millet is high in protein. Little millet is high in fat, comprising majorly of the healthy polyunsaturated fatty acids (PUFA). It contains high phosphorous (220 mg/100 g) and iron (9.3 mg/100 g). It is especially good for people has low body mass (Tables 7.1, 7.2, 7.3 and 7.4 showcases the nutritional composition, essential amino acids profile, vitamins and micronutrient profiling of little millet grains).

7.6 Antioxidant and Bioactive Compounds Present in Little Millet

Antioxidants are compounds that either completely stop or significantly reduce the damage that free radical's unstable molecules produced by the body in response to external stimuli do to cells. Bioactive compounds are chemical compounds found in small amounts in plants as well as certain foods such as fruits, vegetables, nuts, oil, whole grains and they have actions within the body that promote health. Antioxidants such as polyphenols, phenolic compounds, tannins, and flavonoids which are abundant in little millet help to promote health by combating lifestyle diseases like diabetes, cardiovascular disease, obesity, cataracts, cancers, inflammation, and gastrointestinal issues all of which are major issues facing our nation right now. The polyphenols are the biggest group of phytochemicals exhibiting antioxidant, metal

Table 7.1 Nutritional Composition of Little Millet compared to fine cereals per 100 g

Nutrients	Little millet	Rice	Wheat
Proteins (g)	7.7	6.8	11.8
Carbohydrate (g)	67	78.2	71.2
fat (g)	4.7	0.5	1.5
Crude fiber (g)	7.6	0.2	1.2
Mineral matters (g)	1.5	0.6	1.5
Calcium (mg)	17	10	41
Phosphorous (mg)	220	160	306
Iron (g)	9.3	0.7	5.3
Energy (Kcal)	341	345	346

Source: Indirani & Devasena, 2021

Table 7.2 Essential amino acid profile of little millet (mg/g of N)

Essential amino-acids	mg/g of N
Arginine	250
Histidine	120
Lysine	110
Tryptophan	60
Phenyl Alanine	330
Tyrosine	–
Methionine	180
Cystine	90
Threonine	190
Leucine	760
Isoleucine	370
Valine	350

Source: Tomar et al., 2023

chelating, and reducing powers. These antioxidants presented in little millet contribute to health, delay aging, reduce metabolic syndrome and improve the immune system. The bioactive properties of polyphenols include anticarcinogenic, anti-inflammatory, antiviral, and neuroprotective activities (Indirani & Devasena, 2021) (Fig. 7.6 illustrates the antioxidants and bioactive compounds present in little millet).

Little millet (*Panicum sumatrense*) is a nutrient-dense cereal grain that contains a variety of bioactive compounds contributing to its health-promoting properties. These compounds include phenolic acids, flavonoids, and other antioxidants that help combat oxidative stress in the body. The grain is rich in dietary fiber, both soluble and insoluble, which supports digestive health and may aid in managing blood glucose levels. Little millet also contains essential vitamins, particularly B-complex vitamins and vitamin E, as well as important minerals such as iron, zinc, magnesium, and phosphorus. The presence of essential amino acids enhances its nutritional profile, while phytosterols may contribute to cardiovascular health. Additionally, the resistant starch found in little millet acts as a prebiotic, potentially benefiting gut microbiota. These bioactive compounds collectively contribute to the grain's potential in managing chronic conditions like diabetes and cardiovascular diseases, making little millet a valuable addition to a healthy diet (Devi et al., 2014).

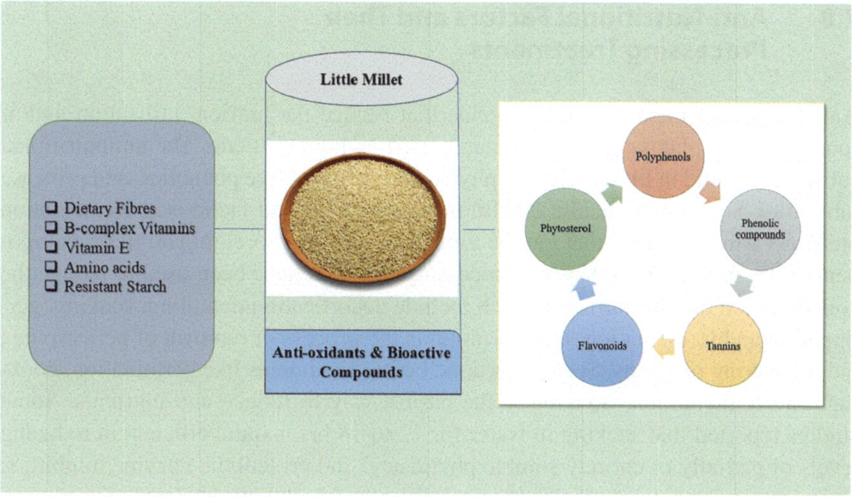

Fig. 7.6 Antioxidants and bioactive compounds

Table 7.3 Vitamin profile of little millet (mg/100 g)

Thiamine	Niacin	Riboflavin	Retinol	Folic acid
0.3	3.2	0.09	0	9.0

Source: Tomar et al., 2023

Table 7.4 Micronutrient profile of millets (mg/100 g)

Mg	Na	K	Cu	Mn	Mb	Zn	Cr	Su	Cl
133	8.1	129	1.00	0.68	0.016	3.7	0.180	149	13

Source: Tomar et al., 2023

7.7 Little Millet as Probiotic and Prebiotic

Probiotic products developed from little millet are a growing field of study in functional foods. Due to its potential to support gut health as a probiotic and prebiotic, little millet has attracted a lot of attention. Little millet is not a probiotic, although it can be used as a substrate for probiotic fermentation. Probiotics are "living micro-organisms" which when administered in adequate amount confers a health benefit on the host (Abd El-Salam et al., 2012). Fermented millet products act as a natural probiotic treatment for diarrhea in young children (Lei et al., 2006). Prebiotics are non-digestible food ingredients that beneficially affect the host by selectively stimulating the growth activity of one or a limited number of bacteria in the colon (Laminu et al., 2011). With a high content of soluble and insoluble fibres, little millet can act as a prebiotic food supply to beneficial gut bacteria, supporting their growth and activity.

7.8 Anti-Nutritional Factors and Their Processing Treatments

Anti-nutritional factors are compounds that reduce the nutrient utilization and/or food intake of plants, or plant products used as human foods. The antinutritional substances found in little millet are phytic acid, tannins, free phenolics, and protease inhibitors. Soaking (in acidic conditions), sprouting, and lactic acid fermentation are processing methods that can be employed in order to eliminate these components (Dey et al., 2022). Several processing methods have been used to reduce the concentration of anti-nutrients which include decortication/dehulling, soaking, germination and fermentation. Decortication is the process of removal of pericarp i.e., outer covering of grains. Soaking is done before treatments like germination, cooking, and fermentation. Soaking is the easiest way to reduce anti-nutrients. Some studies reported that soaking in water for 12 to 18 hrs is quite efficient in reducing levels of partially or entirely soluble phytic acid and proteolytic enzyme inhibitors. Germination is an active phase of metabolism in which anti-nutrients are reduced. Fermentation eliminates anti-nutritional factor. Various microorganisms are used for fermentation, such as Saccharomyces, Lactobacillus, and different types of fungi, yeast, and mold, all of which show different results. Heating includes roasting, boiling, cooking, or autoclaving. Roasting reduced 74.6, 28.4, 98.3, and 97.5% of tannin, phytate, trypsin inhibitor, and protease inhibitor, respectively. However, cooking reduced these by 42, 75.8, 95.8, and 95.8%, respectively (Samtiya et al., 2021) (Fig. 7.7 Illustrates the processing treatments that can be used to eliminate anti-nutritional factors present in little millet).

7.9 Health Benefits of Little Millet

7.9.1 Good Source of Potent Antioxidants

Millet's antioxidants such as polyphenols, phenolic compounds, tannins, flavonoids are not directly related to nourishing the body, but play an important role in promoting health by helping in disease, cataract, cancer, inflammation, and gastrointestinal problems. Polyphenols are considered as "life span essential" due to their role in maintaining health throughout end phase of life.

7.9.2 Act as a Nutraceutical

Nutraceuticals are a class of food-based bio molecules and exibit hypoglycemic, hypocholesteromic, hypolipidmic and anti-glycation properties. Little millet stands out to be one of our "indegenous super food". The case of obesity and diabetes are increasing fast globally. Food containing complex carbohydrates with high fiber and

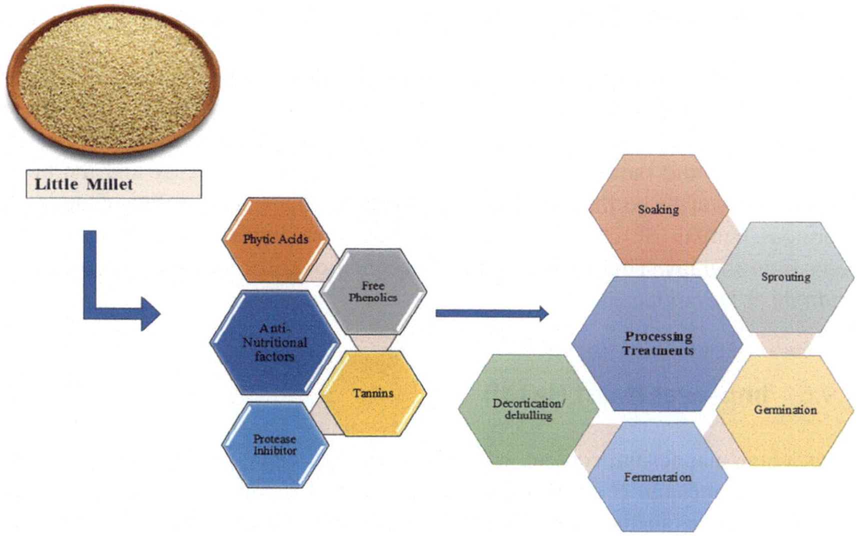

Fig. 7.7 Anti-nutritional factors and their processing treatments

health beneficial phytochemicals has been in demand to combat health issues. Phytates, polyphenols and tannins of millet's food can contribute to antioxidant activity that play important role in health, aging and metabolic disease. Millet can be termed as "food medicine".

7.9.3 Helps Combat Diabetes

High carbohydrate diet can increase plasma glucose, leading to insulin resistance. Little millet is a low glycemic index food, which is a good source of slow digesting carbohydrates and dietary fibre. It takes longer for glucose to enter the blood stream and hence blood sugar levels are stable. This proves to be beneficial for diabetics who have to control rapid rise and decline of glucose in the body.

7.9.4 Cataract Genesis Inhibition

Western countries are facing a major issue of blindness due to retinopathy and cataract worldwide. In diabetes patients there will be an accumulation of sorbitol. This accumulation is mediated by the key enzyme known as aldose reductase. It is present in little millet. This reduces the risk of developing cataracts (Indirani & Devasena, 2021).

7.9.5 Cancer-Fighting

The chief grain of India is millet which is versatile, gluten-free, whole grain of India that is similar to quinoa. Small millet is a nutritional powerhouse and excellent source of protein, fibre, B vitamins, iron, zinc, phosphorous, and magnesium. Chemotherapy and radiation therapy can be exhausting the energy and the body needs the power to go through each treatment. So, these little millets contain an enormous number of antioxidant and phenolic compounds which detoxifies the body from therapy. This grain rectifies the conditions like asthma and bronchitis (Indirani & Devasena, 2021).

7.9.6 Improves Heart Health

It has a high magnesium content, which helps to keep blood pressure and heart rate constant. Magnesium is a mineral that is necessary for many metabolic processes that take place inside the body. This millet-like grain also combats depression, particularly in older people (Indirani & Devasena, 2021) (Fig. 7.8 highlights the health benefits of consuming little millet).

7.9.7 Other Benefits of Little Millet

High amounts of iron help to safe guard many fundamental functions in the body, including general energy and focus, the immune system, gastrointestinal processes, and the regulation of body temperature. Higher amounts of zinc aid in enzymatic reactions, immune function, wound healing, DNA and protein synthesis, and regular growth and development of the body (Antony et al., 1996; McKeown et al., 2002; Chandrasekara & Shahidi, 2010; Lee et al., 2010).

Little millet contains magnesium which can help improve heart health. Vitamin B3 (niacin) in little millet helps lower cholesterol. Little millet is also a good source of phosphorous which helps with fat metabolism, body tissue repair and energy production (Fig. 7.9 shows the diseases that can be cured by consuming little millet).

7.10 Little Millet Based Products

Few recipes which can be prepared using little millet are dosa, idli, pongal, kichadi (Nutritive value of Indian foods, NIN, 2007).

7.10.1 Dosa

Samai dosa is prepared by using one cup of little millet, half cup dhal of black gram, half cup processed rice (puffed), one tablespoon of fenugreek, and a desired amount

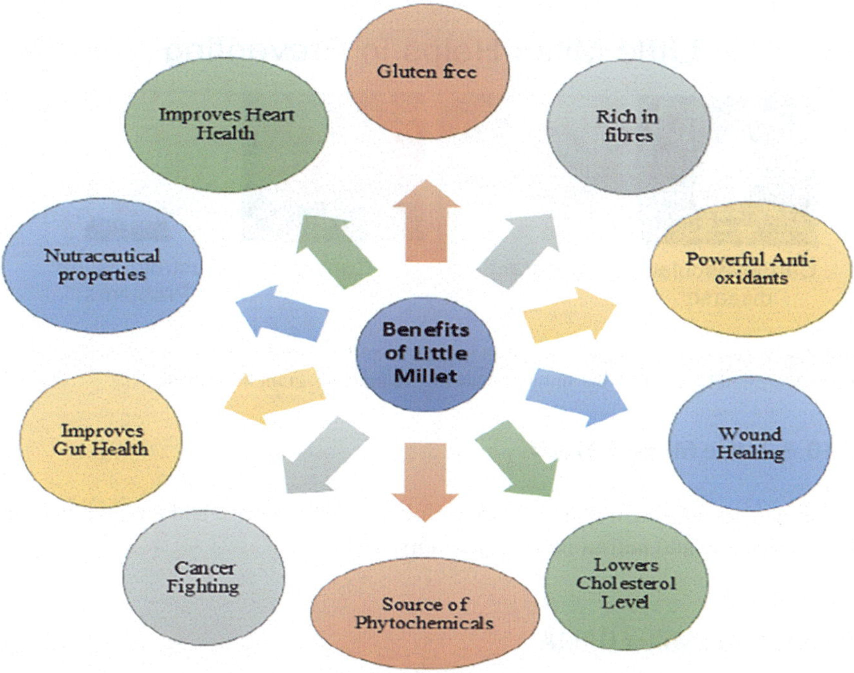

Fig. 7.8 Health benefits of consuming little millet

of salt to taste. Soaked seeds of fenugreek, black gram, and rice are used in the preparation process. All the soaked ingredients are grind properly and kept overnight for fermentation. Mix all the processed ingredients and prepare the dosa (Kumar et al., 2021).

7.10.2 Porridge

Porridge is prepared using the following ingredients; little millet, coconut milk, and salt. Rice is cooked in water and added to coconut milk for 10–15 min, add the desired amount of salt and cool. The porridge is ready to serve (Kumar et al., 2021).

7.10.3 Payasam

Payasam is prepared by using little millet flour, sugar, ghee, milk powder, and milk. The product makes it healthier and tastier by adding fruits before serving (Kumar et al., 2021).

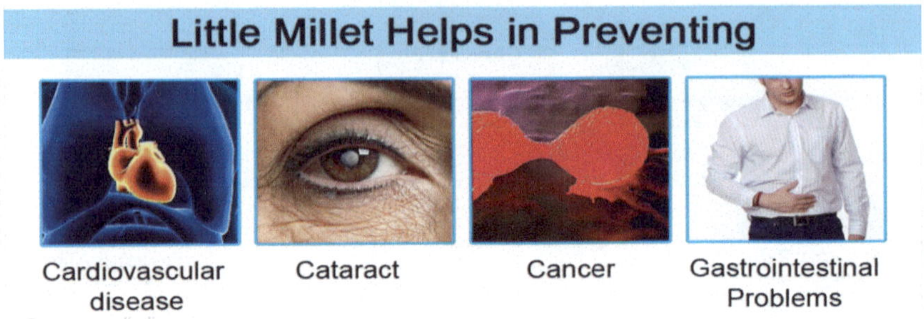

Fig. 7.9 Prevention of disease by consuming little millet. (Source: https://www.medindia.net/dietandnutrition/little-millet-the-immune-boosting-indigenous-grain.htm#about)

7.10.4 Little Millet SALAD

Combine cooked little millet with colourful vegetables, herbs, and a zesty dressing for a refreshing and nutritious salad (Swamy, 2023).

7.10.5 Little Millet UPMA

Prepare a flavourful upma by sautéing onions, vegetables, and spices and then adding cooked little millet (Swamy, 2023).

7.10.6 Little Millet Pulao

Cook little millet with aromatic spices, vegetables, and broth to create a flavourful and wholesome pulao (Swamy, 2023).

7.10.7 Little Millet Kheer

Simmer cooked little millet in milk with sugar, cardamom, and nuts for a delicious and nutritious dessert (Swamy, 2023).

7.10.8 Little Millet Stuffed Bell Peppers

Fill bell peppers with a savoury mixture of cooked little millet, vegetables, and spices, and bake until tender (Swamy, 2023) (Fig. 7.10 variety of foods that can be made using little millet).

Fig. 7.10 Little Millet based food products. (Source: Swamy, 2023)

7.11 Future Perspective of Little Millet

The future outlook for little millet is highly promising in addressing key global challenges such as food security, malnutrition, and climate change impacts. This hardy and nutritious grain is well-positioned to become a cornerstone of sustainable agricultural practices. Innovations in processing methods and the creation of new millet-based products are likely to broaden its market appeal and boost consumer adoption. Little millet's remarkable adaptability to various growing conditions makes it an excellent crop for climate change adaptation strategies in agriculture. Its incorporation into eco-friendly farming systems, combined with favourable policies and growing recognition of its health advantages, could spur more widespread cultivation and use. The grain's capacity to enhance rural economies and maintain crop diversity further highlights its potential significance in shaping future agricultural practices. As the world increasingly turns its attention to underappreciated and underutilized crops, little millet is poised to attract greater interest from the scientific community, policymakers, and the general public. This increased focus could catalyse its transition from a minor crop to a globally significant cereal, playing a vital role in sustainable food systems of the future.

As public awareness of its health benefits grows, little millet is expected to gain prominence in both traditional dishes and contemporary cuisine. Its gluten-free nature positions it well in the growing market for alternative grains. Culinary experts and food technologists are likely to craft novel recipes and products that highlight its adaptability, ranging from gluten-free pastries to convenient snacks and breakfast options. Little millet's impressive nutritional profile, rich in protein, fiber, and essential micronutrients, makes it an ideal crop for functional foods and dietary supplements. Advancements in processing technology may yield improved dehulling and milling techniques, enhancing the texture and taste of millet-based products. The increasing popularity of plant-based diets could spur the creation of little millet protein extracts for use in meat substitutes. Moreover, the grain's potential in brewing gluten-free beer and other fermented drinks may be further investigated. This versatile grain offers a nutritious alternative to conventional cereals, contributing to a more diverse and secure food supply while meeting the growing demand for healthier, environmentally friendly food choices.

Little millet's high fibre content, low glycemic index, and rich nutritional profile make it a potential ally in the fight against noncommunicable diseases like diabetes, obesity, and cardiovascular disorders. Given its prebiotic potential, future research may shed light on its involvement in intestinal health. The presence of antioxidants in small millet may lead to research into its cancer-preventing abilities. Nutritionists may increasingly prescribe little millet as part of a well-balanced diet for weight loss and blood sugar regulation. Its gluten-free status makes it an excellent choice for anyone with celiac disease or gluten intolerance.

7.12 Conclusion

Little Millet One of the minor underutilised millets can play a significant role in food security. The grain is gluten-free and has a glycemic index of 50–65. It can be used as an alternative to cereals like wheat, rice and contains important elements such as protein, carbohydrates, and fat. The grain contains a high concentration of antioxidants and bioactive substances, which can help improve human health.

Consuming little millet can help cure a variety of ailments, including cardiovascular disease, cataracts, cancer, and diabetes.

Little millet, which contains major nutrients including protein, carbohydrates, and fat among others, as well as a wealth of vitamins and minerals, can be utilized as a main food source. Being gluten free and having a low glycemic index little millet can be consumed by persons of all ages. Nutraceuticals such phenolics, gama-amino butyric acid (GABA), lignans, resistant starch, sterols and phytates are also abundant in it. Little millet has a number of health advantages and aids in the treatment of a number of ailments, including cataract, cancer, diabetes, inflammation, and digestive issues. Phytic acid, tannins, free phenolics, and protease inhibitors are a few anti-nutritional components that are a present in little millet. They can be removed through processing methods like sprouting, lactic acid fermentation, and soaking. Little millet has a high nutritional value and can be used in place of grains like rice and wheat. The millet has nutritional benefits and is a strong source of fiber. As many people suffer from conditions such as gluten intolerance and lactose intolerance, it can be used as a replacement for creating a variety of food products such as idli, dosa, khichadi, cookies, payasam, porridges, puffy products, fermented beverages and many more.Conflict of InterestThe authors declare no conflict of interest in th**is work.**

Data Availability Statement Although no datasets have been produced or analysed for this research, data sharing is not applicable.

References

Abd El-Salam, M. H., Hippen, A. R., Salem, M. M., Assem, F. M., & El-Aassar, M. (2012). Survival of probiotic Lactobacillus casei and Enterococcus fecium in Domiati cheese of high conjugated linoleic acid content. *Emiratus Journal of Food and Agriculture, 24*, 98–104.

Adekunle, A. A. (2012). *Agricultural innovation in sub-Saharan Africa: Experiences from multiple-stakeholder approaches*. Forum for Agricultural Research in Africa. ISBN 978-9988-8373-2-4.

Agrawal, P., Singh, B. R., Gajbe, U., Kalambe, M. A., & Bankar, M. (2023). Managing diabetes mellitus with millets: A new solution. *Cureus, 15*, e44908. PubMed: 37814770. https://doi.org/10.7759/cureus.44908

Amadou, I., Gounga, M. E., & Le, G. W. (2013). Millets: Nutritional composition, some health benefits and processing—A review. *Emirates Journal of Food and Agriculture, 25*, 501–508. https://doi.org/10.9755/ejfa.v25i7.12045

Anitha, S., Kane-Potaka, J., Tsusaka, T. W., Botha, R., Rajendran, A., Givens, D. I., Parasannanavar, D. J., Subramaniam, K., Prasad, K. D. V., Vetriventhan, M., & Bhandari, R. K. (2021). A systematic review and meta-analysis of the potential of millets for managing and reducing the risk of developing diabetes mellitus. *Frontiers in Nutrition, 8*, 687428. PubMed: 34395493. https://doi.org/10.3389/fnut.2021.687428

Antony, U., Sripriya, G., & Chandra, T. S. (1996). Effect of fermentation on the primary nutrients in finger millet (Eleusine coracana). *Journal of Agricultural and Food Chemistry, 44*, 2616–2618. https://doi.org/10.1021/jf950787q

Bhat, B. V., Rao, B. D., & Tonapi, V. A. (2018). *The Story of Mil Lets*. ICAR-Indian Institute of Millets Research.

Chandrasekara, A., & Shahidi, F. (2010). Content of insoluble bound phenolics in millets and their contribution to antioxidant capacity. *Journal of Agricultural and Food Chemistry, 58*, 6706–6714. PubMed: 20465288. https://doi.org/10.1021/jf100868b

Crops. (n.d.). Retrieved from https://www.crops.org/news/science-news/big-potential-little-millet/

Devi, P. B., Vijayabharathi, R., Sathyabama, S., Malleshi, N. G., & Priyadarisini, V. B. (2014). Health benefits of finger millet (*Eleusine coracana* L.) polyphenols and dietary fiber: A review. *Journal of Food Science and Technology, 51*, 1021–1040. PubMed: 24876635. https://doi.org/10.1007/s13197-011-0584-9

Dey, S., Saxena, A., Kumar, Y., Maity, T., & Tarafdar, A. (2022). Understanding the antinutritional factors and bioactive compounds of Kodo Millet (*Paspalum scrobiculatum*) and Little Millet (*Panicum sumatrense*). *Journal of Food Quality, 2022*, 1–19. https://doi.org/10.1155/2022/1578448

FAO. (2009). FAOSTAT. *Food and Agriculture Organization of the United Nations. FAOSTAT*. Retrieved from faostat.fao.org/site/339/default.aspx

Ganapathy, K. N. (2017). Genetic improvement in little millet. In *Millets and Sorghum* (pp. 170–183).

Gautam, N., Singh, M., Arya, P., Kumar, R., Bhargava, P., & Gupta, A. (2023). *Goodness of millets in daily meals through easy & healthy recipes* (1st ed.). Bundelkhand University. ISBN 978-81-958954-8-9.

Gayathri, A. (2015). *Little Millet Nutrition, Health Benefits and Facts You Must Know*

Gopalan, C., Ramasastri, B. V., & Balaubramanian, S. C. (2010). *Nutritive value of Indian foods*. National Institute of Nutrition, Indian Council of Medical Research.

Goron, T. L., & Raizada, M. N. (2015). Genetic diversity and genomic resources available for the smallmillet crops to accelerate a New Green Revolution. *Frontiers in Plant Science, 6*(1), 57. https://doi.org/10.3389/fpls.2015.00157

Guha, M., Sreerama, Y. N., & Malleshi, N. G. (2015). Influence of processing on nutraceuticals of Little Millet (Panicum sumatrense). In *Processing and impact on active components in food* (pp. 353–360). Elsevier. https://doi.org/10.1016/B978-0-12-404699-3.00042-1

Hegde, P. S., & Chandra, T. S. (2005). ESR spectroscopic study reveals higher free radical quenching potential in kodo millet (*Paspalum scrobiculatum*) compared to other millets. *Food Chemistry, 92*, 177–182. https://doi.org/10.1016/j.foodchem.2004.08.002

Hiremath, S. C., Patil, G. N. V., & Salimath, S. S. (1990). Genome homology and origin of *Panicum sumatrense* (Gramineae). *Cytologia, 55*, 315–319. https://doi.org/10.1508/cytologia.55.315

ICRISAT. (2007). *2007 Annual report*. International Crops Research Institute for the Semi-Arid Tropics. Retrieved from https://test1.icrisat.org/Publications/EBooksOnlinePublications/AnnualReport-2007.pdf

Indirani, K., & Devasena, M. (2021). Review on nutritional profiles and health benefits of little millets-India. *International Journal of Research in Engineering and Science (IJRES), 9*(11), 07–11.

Johnson, M., Deshpande, S., Vetriventhan, M., Upadhyaya, H. D., & Wallace, J. G. (2019). Genome-wide population structure analyses of three minor millets: Kodo millet, little millet and proso millet. *Plant Genome, 12*, 1–9. https://doi.org/10.3835/plantgenome2019.03.0021

Kumar, P., Prasad, R., & Kumar, R. (2016). Underutilised Millets: A way to nutritional security. *International Journal of Agticulture and Environment Research, 4*, 726–734.

Kumar, A., Tripathi, M. K., Joshi, D., & Kumar, V. (2021). Millets and millet technology. In *Millets and millet technology*. Springer. https://doi.org/10.1007/978-981-16-0676-2

Kumari, A. N., Salini, K., & Veerabadhiran, P. (2010). Morphological characterization and evaluation of little millet (*Panicum sumatrense* Roth. ex. Roem. and Schultz.) germplasm. *Electronic Journal of Plant Breeding, 1*, 148–155.

Kushwaha, A. S., Sirohi, R., & Tarafdar, A. (2018). Effect of hydrothermal treatment and milling parameters on milling and nutritional qualities of finger millet (*Eleusine coracana*). *Journal of Agricultural Engineering, 55*(4), 34–46.

Ladumore, V., Jumabhai, D. G., & Patel, S. N. (2021). Genetic diversity studies in little millet (Panicum sumatrense Roth. Ex. Roemer and Schultes) genotypes:A multivariatePCA analysis. *Pharma Innovation Journal, 10*, 1067–1071.

Laminu, H. H., Modu, S., & Numan, A. I. (2011). Production, in vitro protein digestibility, phytate content and acceptability of weaning foods prepared from pearl millet (Pennisetum typhoideum) and cowpea (*Vigna unguiculata*). *International Journal of Nutrition and Metabolism, 3*, 109–113.

Lee, S. H., Chung, I. M., Cha, Y. S., & Park, Y. (2010). Millet consumption decreased serum concentration of triglyceride and C-reactive protein but not oxidative status in hyperlipidemic rats. *Nutrition Research, 30*, 290–296. https://doi.org/10.1016/j.nutres.2010.04.007

Lei, V., Friis, H., & Michaelsen, K. F. (2006). Spontaneously fermented finger millet product as a natural probiotic treatment for diarrhea in young children: An intervention study in Northern Ghana. *International Journal of Food Microbiology, 110*, 246–253. https://doi.org/10.1016/j.ijfoodmicro.2006.04.022

McKeown, N. M., Meigs, J. B., Liu, S., Wilson, P. W., Jacques, P. F. (2002). Whole- grain intake is favorably associated with metabolic risk factors for type 2 diabetes and cardiovascular disease in the Framingham Offspring Study, *American Journal of Clinical Nutrition, 76*(2), 390–398.

Medindia. (n.d.). Retrieved from https://www.medindia.net/dietandnutrition/little-millet-the-immune-boosting-indigenous-grain.htm#about

Neeharika, B., Suneetha, J., Kumari, A., & Tejashree, M. (2020). Sensory profiling of germinated Little Millet at different incubation times. *Chemical Science Review and Letters, 33*, 121–124. https://doi.org/10.37273/chesci.CS205101111

Nirmalakumari, A., Salini, K., & Veerabadhiran, P. (2010). Morphological characterization and evaluation of Little millet (*Panicum sumatrense* Roth. ex. Roem. and Schultz.) Germplasm. *Electronic Journal of Plant Breeding, 1*.

NROOTSFOOD. (n.d.). Retrieved from https://nrootsfood.com/product/little-millet-grains/

Nutritive value of Indian Foods, NIN. (2007). *Millet in Your Meals*. Retrieved from www.sahajasamrudha.org/

Obafaye, R. O., & Omoba, O. S. (2018). Orange peel flour: A potential source of antioxidant and dietary fiber in pearl-millet biscuit. *Journal of Food Biochemistry, 42*, e12523. https://doi.org/10.1111/jfbc.12523

Reddy, O. S. K. (2017). *Smart millet and human health*. Green Universe Environmental Services Society.

Saloni, S., Sindhu, Sujata, , Kumari, S. & Sandhya, S. (2018) Little millets: Properties, functions and future prospects. International Journal of Agricultural Engineering, 11, 179–181. https://doi.org/10.15740/HAS/IJAE/11.Sp

Samtiya, M., Soni, K., Chawla, S., Poonia, A., Sehgal, S., & Dhewa, T. (2021). Key antinutrients of millet and their reduction strategies: An overview. *Acta Scientifci Nutritional Health, 5*, 68–80. https://doi.org/10.31080/ASNH.2020.05.0963

Sivakumar, S., Franco, O. L., Thayumanavan, B., Murad, A. M., Manickam, A., Mohan, M., & Mridula, M. (2006). Cloning and structural analysis of an Indian little millet (Panicum sumatrense) zein-like storage protein: Implications for molecular assembly. *Biochemistry Biokhimiia, 71*, 1183–1191. https://doi.org/10.1134/s0006297906110034

Swamy, K. R. M. (2023). Origin, domestication, taxonomy, botanical description, genetics and cytogenetics, genetic diversity and breeding of little millet (*Panicum sumatrense* Roth. ex. Roem. & Schult.). *International Journal of Current Research, 15*, 26170–26192.

The Hindu. (2015). The Journey of Little millet. *Hindu*

Tiwari, A., Kesarwani, K., Sharma, A., Ghosh, T., Bisht, N., & Punetha, S. (2022). Drought stress in millets and its response mechanism. In *Advances in plant defense mechanisms*. IntechOpen eBooks. https://doi.org/10.5772/intechopen.105942

Tomar, A., Mishra, A. K., Singh, S. P., Khan, H. H., Patel, P., & Singh, R. P. (2023). Nutritional benefit of small millets: Food security & sustainability in India. *Journal of Pharmaceutical Innovation, 12*(2), 3028–3033.

Wikipedia. (2023). Panicum sumatrense. *From Wikipedia, the Free Encyclopedia*. Retrieved from https://en.wikipedia.org/wiki/Panicum_sumatrense

8 Sustaining Livelihood with Millet Agriculture

Shubhangi Nigam, Jaspreet Kaur, and Anushka Sharma

Abstract

The sustainable and resilient nature of millet agriculture as a means to support livelihoods in diverse Communities. Millets, a group of small-seeded, hardy cereal crops, have been cultivated for centuries Across various regions, offering a multitude of benefits that contribute to the overall well-being of Communities. Millet agriculture stands out as a resilient solution in the face of climate change, Demonstrating adaptability to diverse environmental conditions. Furthermore, millets are rich in Nutritional value, providing essential nutrients such as protein, fiber, and micronutrients. The Cultivation of millets Supports food security by offering a diverse and nutritious dietary base. The Socio-economic impact of millet agriculture extends beyond the farm, fostering community Resilience. The cultivation of millets can empower local farmers, especially in marginalized areas, By providing income opportunities and reducing dependency on external resources. Millet Agriculture emerges as a sustainable and resilient strategy for sustaining livelihoods. Its adaptability, Nutritional value, and socio-economic benefits position millets as a key player in fostering Community well-being and resilience in the ever changing global landscape. This paper aims to shed Light on the potential of millet agriculture as a transformative force in promoting sustainable Livelihoods and enhancing the overall quality of life for diverse communities.

Keywords

Sustainability · Entrepreneurship · Socio-economic impact · Ecological balance of millets · Agriculture diversification

S. Nigam (✉) · J. Kaur · A. Sharma
Department of Food Technology, Institute of Engineering and Technology, Bundelkhand University, Jhansi, Uttar Pradesh, India

C. B. Singh et al. (eds.), *Global Millets Production for a Sustainable Future*,
https://doi.org/10.1007/978-3-031-88157-2_8

Keywords

Sustainability: Sustainability is a social goal for people to co-exist on earth over a long time. Sustainability usually has three dimensions: environmental, economic, and social.

Entrepreneurship: Entrepreneurship is the creation or extraction of economic value in ways that generally entail beyond the minimal amount of risk, and potentially involving values besides simply economic ones.

Socio-economic impact: To study how economic activity affects and is shaped by social processes. In general, it analyzes how modern societies progress, stagnate, or regress Because of their local or regional economy, or the global economy.

Ecological balance of millets: Millets are versatile, need 70% less water than rice to Grow, take half as long to mature as wheat, and use 40% less energy to process. They offer a Comprehensive answer for sustained food security in the face of drought, water shortage And climate change. Millets play a crucial role in enhancing soil health by promoting better Water retention, reducing soil erosion, and fostering nutrient cycling. Their deep root Systems and efficient resource use contribute to improved soil structure and fertility.

Agriculture diversification: Agricultural diversification occurs when more species, plant Varieties or animal breeds are added to a given farm or farming community, and this may Include landscape diversification—different crops and cropping systems interspersed in Space and time

8.1 Introduction

Millet agriculture is a viable way to sustain livelihoods with several environmental, Nutritional, and economic advantages. Pearl millet, sorghum, and finger millet are Examples of tough crops that can withstand a wide range of environmental conditions. Adopting millet cultivation promotes community economic well-being in addition to Ensuring food security. Unlike other crops, millets may flourish in dry and semi-arid Locations due to their resistance to drought. This increases their resilience against the Effects of climate change by providing farmers with a steady stream of income in the face Of erratic weather patterns. Because of their high vitamin content, millet makes a Wholesome dietary source. Communities can battle malnutrition and related health issues By encouraging the intake of millet. This change to a diversified, nutrient-dense diet helps To Enhanced general well-being. Furthermore, millets are grown to support sustainable arming methods and biodiversity. In contrast to monoculture, millet agriculture lessens The need for chemical inputs while promoting ecological balance. Thus, long-term Sustainability is encouraged and soil fertility is maintained. There are market opportunities For millet producers due to the rising interest in ancient grains around the world and the Growing need for gluten-free substitutes. Using millets to diversify agricultural techniques Creates opportunities for local entrepreneurship and revenue generating. Millet agriculture Is a multidimensional strategy that tackles food security,

Fig. 8.1 Millet farming in India. Source: https://images.app.goo.gl/8TGq5QbeYCtckizZ6

nutrition, environmental Sustainability, and economic growth in order to sustain livelihoods. Uplifting Adopting Millet farming by farmers helps to create resilient and prosperous communities while also Securing their livelihoods (Fig. 8.1).

8.2 Millet Crops Overview

Overview of Millet Crops A collection of small-seeded grains from the Poaceae family is Referred to as millet. These ancient grains are well known for their tenacity, adaptability, And nutritional content and have been grown for thousands of years. There are numerous Varieties of millets, each with special qualities appropriate for particular temperatures and Geographical areas.

8.2.1 Pearl Millet

A common crop in arid and semi-arid areas, pearl millet is prized for its Exceptional resistance to heat and drought. Its structure is grassy and tall, and its Inflorescences are spike-like.

Pearl millet is an important source of vitamins, protein, and energy due to its high Nutritional content.

8.2.2 Sorghum

Sorghum is a multipurpose millet that can be used for industrial, culinary, And fodder purposes. It is an essential crop in a variety of climates since it adapts well to Regions with high and low rainfall.

Grain sorghum is devoid of gluten and a valuable source of fiber, antioxidants, and carbs.

8.2.3 Finger Millet

Cultivated mostly in South Asia and Africa, finger millet is also referred to as ragi.

It is a nutrient-dense option because it is high in calcium, iron, and other minerals. The crop's low water requirements make it ideal for small-scale and subsistence farming.

8.2.4 Millet Foxtail

Foxtail millet is becoming more and more popular as a grain and in a Variety of culinary applications due to its gluten-free status.

Because it matures quickly, farmers can harvest it numerous times a year, making it a Dependable crop.

8.2.5 Proso Millet

It is grown in both arid and temperate climates due to its adaptability to a Wide range of weather conditions. It has a short growing season and is typically used for Grain, fodder, and birdseed. Millet crops are recognized not only for their tolerance to varied Environmental circumstances but also for their contribution to sustainable agriculture. Millets, being climate-smart crops, are essential for maintaining food security, encouraging Biodiversity, and bolstering the livelihoods of people worldwide (Fig. 8.2).

8.3 Nutritional and Health Benefits of Millets

New research has shed light on the potential use of millets in therapeutic diets. Millets also contain several bioactive phytochemicals including feraxans, Lignans, ß-glucan, inulin, resistant starch, sterols and phenolic compounds (e.g., ferulic acid, caffeic acid and quercetin). Studies have supported the role Of polyphenols in antioxidant, anti-inflammatory, antiviral and neuroprotective Activities, which have shown to be beneficial against cardiovascular disease, Diabetes, high blood pressure, high cholesterol, inflammatory diseases, Metabolic syndrome. Here is how millets can be therapeutic in certain lifestyle Diseases.

Fig. 8.2 Different types of millets in India. Source: https://images.app.goo.gl/mdJMcXLrhUwEzvj48

8.3.1 Diabetes

India has the second highest prevalence of diabetes only after China with Approximately 77 million Indians affected by it. Lifestyle modifications with Dietary intervention is key in managing type 2 diabetes. The high fibre content in Millets aid in slow release of sugars into the blood stream maintaining a low Post prandial glucose level. The B vitamins in millets support better Carbohydrate utilization and ultimately millets serve as a low glycaemic food, Especially effective for type 2 diabetes.

8.3.2 Cardiovascular Disease

According to the Indian Council of Medical Research and Registrar General of India, India accounts for almost 60% of the world's heart disease burden. Millets contain significant amounts of Magnesium, known to reduce the effects of heart attacks. The antioxidant Nature of millets are known to help reduce oxidative stress and the fibre in them Helps minimize the accumulation of bad cholesterol (LDL) and increase good Cholesterol (HDL). Fibre in millets also aid in delaying the absorption of Cholesterol, dietary fat and glucose, thus contributing to heart health.

8.3.3 Hypertension

In India, 24% of men and 21% of women over the age of 15 have hypertension, according to The National Family Health Survey-5. One of the risk factors for heart disease and stroke is High blood pressure. Dietary Approaches to Stop Hypertension, or DASH diet, encourages The eating of whole foods high in fiber. Studies indicate that consuming millet may help Manage hypertension. Millets are high in dietary fiber, and they also include minerals like Calcium and magnesium, as well as certain antioxidants, which help to keep blood Pressure levels within normal ranges. They guarantee a healthy serum lipid profile as well.

8.3.4 Celiac Disease

Individuals with gastrointestinal diseases such as celiac disease are known to be gluten Intolerant. The family of gluten proteins that cause inflammation on the inside lining of the Intestines in gluten intolerant individuals includes gliadin found in wheat, horedein in Barley, scaling in rye, and venin found in oats. Because they are inherently free of gluten, Millets are a great alternative to cereal grains for people with celiac disease.

8.4 Millets Cultivation in Different Regions of India

8.4.1 Overview

The world's largest producer of millet is India. In 2020, the two millets grown in India, Pearl Millet (Bajra) and Sorghum (Jowar), would account for roughly 19% of global production. In 2020, the world's production of millets was made up of 40.51% Pearl Millet from India and 8.09% Sorghum. The states of Rajasthan, Karnataka, Maharashtra, Uttar Pradesh, Haryana, Gujarat, Madhya Pradesh, Tamil Nadu, Andhra Pradesh, and Uttarakhand are the main producers of millets in India. Together, these ten states currently Produce 98% of all millets produced in India between 2020 and 2021. Rajasthan, Karnataka, Maharashtra, Uttar Pradesh, Haryana, and Gujarat are the six states that account for the.

Majority of the 83% of the millet produced worldwide. Twenty-eight percent of India's Millet is produced in Rajasthan. India produces a wide range of millets, including pseudo Millets like buckwheat and amaranths, as well as Pearl Millets, Sorghum, Finger Millet, Foxtail, Kodo, Barnyard, Proso, and Little Millet. The majority of India's millets are produced As finger millet (Ragi), sorghum (Jowar), and pearl millet (Bajra).

8.5 Historical Importance of Millet Crops in Indian Agriculture

8.5.1 General

Indian agriculture has benefited much from the use of millet crops, which have Contributed greatly to the nation's long agricultural history. Cultivated for generations, These hardy, small-seeded grains have shaped not only the agricultural landscape but also Cultural and dietary customs. India has a long history of millet farming; evidence points to Its use as early as 2500 BCE in the Harappan culture. Millets such as finger millet (ragi), Sorghum (jowar), pearl millet (bajra), and foxtail millet have been staple crops in many Parts of the country for millennia. They are essential because of their capacity to adapt to a Wide range of climatic conditions, particularly in dry and semi-arid areas where other crops Are difficult to grow. One of the most important millet historical aspects cultivation in India Is its role in ensuring food security during challenging times. Millets are known for their Drought resistance and ability to thrive in low-fertility soils. This resilience has been crucial During periods of erratic rainfall, allowing farmers to sustain agricultural productivity even In adverse conditions. In addition to their agricultural significance, millets have deep Cultural roots in India. They have been an integral part of traditional diets, particularly in Regions where rice and wheat may not be as prevalent. Millets are often used in the Preparation of various traditional dishes, providing essential nutrients and dietary diversity. The cultural importance of millets is reflected in rituals, festivals, and local culinary Practices, enriching the culinary heritage of different communities The Green Revolution in The mid-twentieth century that prioritized high-yielding rice and wheat cultivars, overshadowing. These traditional crops and causing a fall in millet farming. But because of their nutritive Value and environmental sustainability, millets have attracted attention again in recent Years. Millets are a nutritious option for a balanced diet because they are high in fiber, Minerals, and have a low glycemic index. Moreover, millets are cultivated in accordance With sustainable agricultural methods. Compared to rice and wheat, these crops require Less inputs, such as water and synthetic fertilizers, which helps to conserve resources and Lessens the impact of agriculture on the environment.

8.5.2 Role of Government

The Indian government has taken steps to encourage the production of millets since it Understands their importance. To increase the production of millet, improve the livelihoods Of farmers, and contribute to nutritional security, initiatives such as the National Food Security Mission for Millets have been implemented.

Fig. 8.3 Global status of millet production. Source: https://images.app.goo.gl/SvvbYeJegmNzzeUT7

8.5.3 Conclusion

In conclusion, millet crops have played a variety of roles in Indian agriculture throughout History. Millets have been an essential part of the nation's agricultural and cultural Landscape for ages, both because of their durability in harsh environments and their Historical farming methods. The resurgence of interest in millets highlights their ongoing Significance in the changing face of Indian agriculture and shows a greater understanding Of their nutritional and environmental advantages (Fig. 8.3).

8.6 Economic and Social Benefits

Worldwide, millets are being heralded as "super crops." Although millets have been Produced and consumed for generations, the Food and Agriculture Organization (FAO) has Only recently declared 2023 to be the "Year of the Millets" based on India's request. In light Of global food insecurity, malnutrition, agricultural distress, and climate change, millet Production is beginning to emerge as a viable alternative. Around 700 million people Globally suffer from hunger, with African and Asian nations being the most severely Afflicted. Problems with food security have been felt on every continent. Millets aid in Addressing these shortages and have the potential to become viable alternatives for Building healthy dietary patterns due to their high nutritional profiles. Numerous areas, Such as the Middle East and Africa have been bringing the declining water levels to the Attention of the world. The strain that present staple crops place on the water table could Be reduced by cultivating water-efficient crops like millets in these regions. Making the Case for millet cultivation is pertinent in the current environment, where the main goals are Creative land-use patterns, more revenue opportunities for farmers, and healthier options For consumers.

8.6.1 Increasing Global Demand

With $201 million in gross trade, millets ranked 3641st among all products worldwide in 2019. The demand for millets has increased significantly over the past few years, as seen By a 44.8% increase in global shipments between 2018 and 2019. India is the world's Largest producer of these grains, but the United States accounts for the majority of Worldwide exports with a share of about 30%. Asia Pacific, the Middle East, and the EU are The main locations of worldwide demand for millets, with 50% of the world's millets Produced being imported into these regions, along with Indonesia, Germany, Iran, Belgium, And South Korea. Although millets are mostly consumed, a small percentage (7%) are also Used as fuel and animal feed.

8.6.2 Emergence of the Asia-Pacific in the Global World

By 2025, the millet market is expected to reach a value of over 12 billion USD. Because of Its significant volume share in international trade, pearl millets are one of the most sought-After commodities, and the Asia-Pacific region is quickly becoming a hub for millet activity. India is a major participant in terms of its capacity to meet worldwide demand for millets, Accounting for close to 20% of the crop's global export value. In addition, China is gradually Increasing its millet yields, which is not surprising given that millet has been grown there for Generations. The United Arab Emirates, Nepal, Saudi Arabia, and Germany are the biggest Export destinations. Over a half-century of growth has occurred in the amount of millets Sent to Nepal and the United Arab Emirates, suggesting a growing market for these crops. Kenya and Oman, two nations whose shares of India's millets exports are comparatively Smaller, have seen even greater increases in their export volumes in recent years. Millets as Productive crops in areas with limited resources.

Rain-fed farming environments, low input requirements, and local knowledge within Farming communities increase the value of millets as a crop in developing nations with Comparable climates. With rice and wheat taking center stage, millets account for less Than 2% of the world's cereal consumption. The previous few decades' agricultural Practices have put pressure on water supplies and altered the carrying capacity Restrictions on the land that is currently in use. Because of this, millets are regaining Popularity as staple grains in many of the semi-arid tropical regions of Africa and India, Where the production of other important food crops is hampered by low precipitation and Declining soil quality. Nearly 38% of the country's domestic produce is consumed in India, A significant customer.

There has been a slow resurgence of interest in millets, mostly in areas where they were Traditionally used as a staple crop. It is motivated by a resurgence of interest in their health Advantages for a population living in cities with changing dietary needs. The acceptance of Organically grown ancient grains—cereals that have experienced little to no genetic change Over time—is trending worldwide in tandem with this increasing movement. To address this Requirement, millets produced with as little inorganic input as possible are the best option. Products made

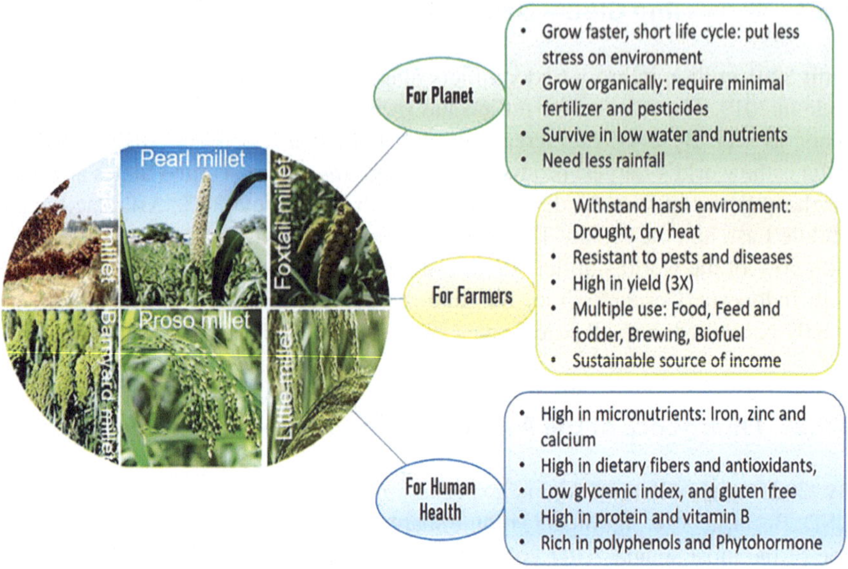

Fig. 8.4 Mainstreaming asphan millet for advancing climate smart agriculture to secure nutrition and health. Source: https://images.app.goo.gl/ETp59a1BqLJsmYrh9

from millet have also become popular choices for those following plant Based or gluten-free diets. Two main areas where millets are being considered as viable Substitutes are breakfast cereals and plant-based dairy products (Fig. 8.4).

8.7 Challenges in Millet Production

Numerous obstacles affect millet cultivation and agriculture, which affects farmers as well as the security of food supply worldwide. A family of small-seeded grasses known as millets are important crops in many areas because they offer necessary nutrients and give Farmers a robust crop to grow in a variety of climates. Nevertheless, a number of Challenges impede their growth and output.

8.7.1 Climate Variability

Growing climate pattern uncertainty is one of the main obstacles to millet production. Millets are frequently cultivated in dry and semi-arid areas, where they are seriously Threatened by sudden changes in rainfall and high temperatures. These problems are made Worse by climate change, which results in longer droughts, changes in crop seasons, and a Shortage of water. To deal with these erratic circumstances, which may have an impact on Agricultural production, farmers must modify their methods (Fig. 8.5).

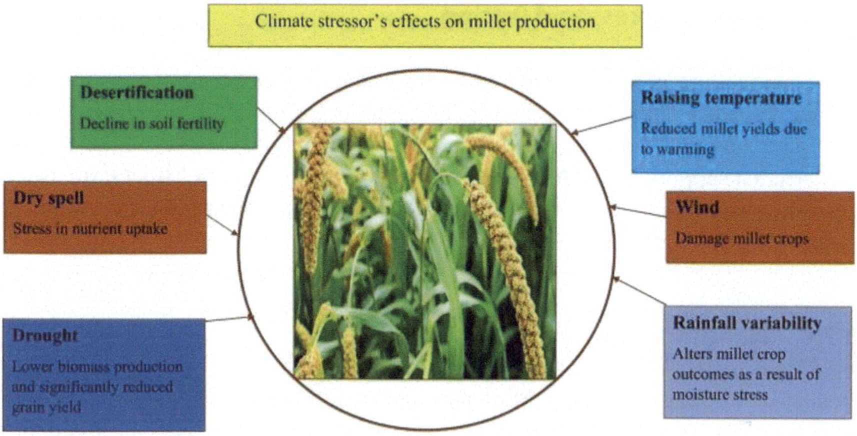

Fig. 8.5 Climate stressors effect on millet production. Source: https://images.app.goo.gl/yyt44SjAHS8ccXnY6

8.7.2 Limited Research and Development

Millets receive a significantly smaller amount of Research and development attention than other cereal crops like rice, wheat, and maize. The creation of robust and high-yielding millet variants is hampered by a lack of funding. To Increase yield, better seeds, pest-resistant strains, and agronomic techniques unique to Millet farming are necessary. The advancement required to sufficiently address these Issues is hampered by a lack of acknowledgment and financing.

8.7.3 Pest and Disease Management

A variety of pests and diseases can severely lower The yields of millet crops. The issue is made worse by poor pest management techniques And restricted access to efficient pesticides. Reducing these risks and guaranteeing Sustainable millet production requires creating integrated pest management plans, Encouraging crop rotation, and improving farmers' understanding of pest and disease Management.

8.7.4 Lack of Infrastructure

Inadequate infrastructure is a major problem in many areas Where millets are farmed. Post-harvest losses arise from farmers' struggles to get their Produce to markets due to inadequate transportation infrastructure. Inadequate storage Facilities further increase the risk of spoiling millets and lower their market value. Ensuring That the advantages of millet production reach rural regions requires improving the Infrastructure in these areas.

8.7.5 Market Access and Pricing

Millet agriculture is further hindered by restricted market Access and unattractive pricing systems. There are frequently less options for millet Producers due to the dominance of larger, more conventional crops in international Markets. Encouraging policies that support equitable pricing and market integration for Millet farmers are critical to maintaining their means of subsistence and advancing the Development of these healthful grains.

8.7.6 Hanging Dietary Preferences

Demand for millet may change if dietary habits shift in favor of more widely consumed cereals. Traditional and regionally consumed millets may become less popular as Urbanization and globalization affect eating habits. Countering these trends and ensuring The continuing cultivation of these essential crops can be achieved by raising awareness of The nutritional benefits of millets and including them into a variety of diets. A comprehensive strategy that includes climate-resilient farming methods, more R&D, Better infrastructure, and supporting legislation is needed to address the issues facing Millet production. To overcome these challenges and guarantee a resilient future for millet Farmers and their communities, it is imperative to acknowledge the significance of millets In attaining global food security and promoting sustainable agricultural methods.

8.8 Strategies to Promote Millet Agriculture

In order to ensure sustainable agriculture, improve nutrition, and increase food security, Millets agriculture must be encouraged. Sorghum, finger millet, pearl millet, and foxtail Millet are examples of hardy crops that have several advantages. The following techniques Can be put into practice to help millets agriculture become more widely adopted (Fig. 8.6).

8.8.1 Awareness Campaigns

Launch extensive awareness campaigns to inform farmers, consumers, and decision-Makers about the advantages of millets for the environment and nutrition. Draw attention To their great nutritional content, low water requirements, and resistance to bad weather.

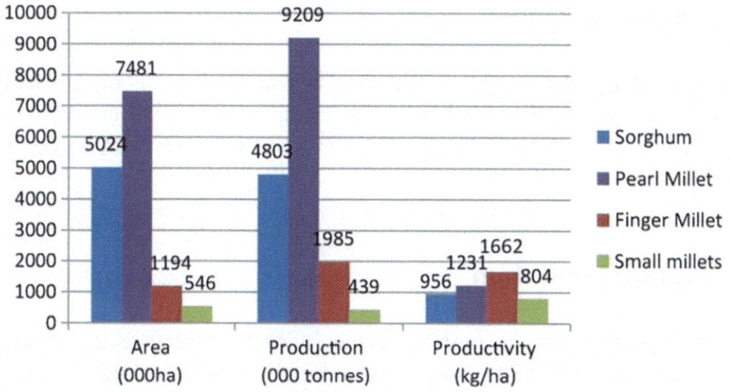

Fig. 8.6 Rate of different millet production. Source: https://www.google.com/imgres?imgurl=https%3A%2F%2Fmedia.springernature.com%2Flw685%2Fspringer-static%2Fimage%2Fchp%253A10.1007%252F978-981-16-7224-8_2%2FMediaObjects%2F482871_1_En_2_Fig1_HTML.png&tbnid=xm1_97fx8FP_OM&vet=1&imgrefurl=https%3A%2F%2Flink.springer.com%2Fchapter%2F10.1007%2F978-981-16-7224-8_2&docid=5Jg-GVyO4mhlZM&w=685&h=356&itg=1&hl=en-US&source=sh%2Fx%2Fim%2Fm1%2F4&kgs=eca89d1e82a8f12b&shem=abme%2Ctrie

8.8.2 Research and Development

Spend money on research to create disease-resistant and high-yielding millet cultivars. To Hasten the creation and spread of better millet varieties, promote cooperation between Universities, business sector organizations, and agricultural research centers.

8.8.3 Financial Incentives

Offer millets farmers financial incentives such grants, subsidies, And low-interest loans. By reducing the initial investment expenses, these incentives can Increase the appeal and viability of millet cultivation from an economic standpoint.

8.8.4 Infrastructure Development

Upgrade the networks of transportation, storage, and Irrigation systems that make up agriculture's infrastructure. Improving infrastructure will Make it easier to distribute millet goods to markets efficiently and assist lower post-harvest Losses.

8.8.5 Market Linkages

Create a network of retailers, processors, and export markets to help Millet farmers build strong market ties. Establishing a stable market for millet goods Guarantees farmers a consistent revenue stream, thereby incentivizing them to persist with Millet farming.

8.8.6 Dietary Promotion

Work with health groups and nutritionists to highlight millets as a Nutrient-dense dietary choice. Create dietary guidelines that highlight millets Contributions to preventing malnutrition and enhancing general health.

8.8.7 Policy Support

Create and carry out regional and national policies that are helpful. This could include millet-promoting policies pertaining to procurement, price, and land-use planning.

8.8.8 Building

Provide farmers with training courses to improve their expertise in millet Farming methods. To increase total productivity, equip them with sustainable agriculture, Organic farming techniques, and modern agricultural technologies.

8.8.9 Community Engagemen

Utilize self-help organizations and farmer to promote community involvement. To develop the millet farming Community, promote information exchange, group purchases of inputs, and cooperative Marketing initiatives.

8.8.10 School Feeding Programs

Include millets in school feeding programs to establish a Predilection and early awareness of millet-based meals. This creates a market for millet Goods in addition to promoting nutrition in youngsters.

8.8.11 Climate Smart Agriculture

Because millets may grow in a variety of agro-ecological Zones, emphasize them as a climate-smart crop. Stress their contribution to climate Change adaptation and sustainable farming methods.

8.8.12 International Cooperation

Exchange information, technology, and best practices for Millet growing with foreign organizations and nations. Accelerating development can be Achieved by worldwide learning from effective millet marketing methods.

8.9 Suggestion to Promote Millet Agriculture

In order to address global issues like malnutrition and climate change, improve rural Livelihoods, and create sustainable food systems, millets agriculture must be encouraged.

The following are detailed recommendations to improve millets agriculture.

8.9.1 Diversification Incentives

Farmer diversification to include millets should be Encouraged by governments and agricultural organizations. This can take the kind of Financial assistance for millet cultivation's early phases as well as subsidies for seeds, Tools, and other inputs.

8.9.2 Development and Research

Make large research and development investments to improve millet varieties. This covers Breeding for increased resilience to the climate, insect resistance, and yield. Creating Research collaborations with corporate companies and agricultural colleges can hasten the Creation of high-yield millet.

8.9.3 Extension Services

To inform farmers about contemporary millet farming methods, Strengthen extension services. To spread knowledge on the best ways to plant, manage Water, and control pests, training sessions, workshops, and demonstrations can be Arranged.

8.9.4 Water Management Practice

Although millet is known to withstand drought, Effective water management techniques can increase its resistance even further. To Maximize water utilization in millet farming, stimulate rainwater gathering, employ water-Efficient irrigation techniques, and support the adoption of technology like drip irrigation.

8.9.5 Market Access

By creating direct connections between producers and consumers, you Can increase the market access available to millet farmers. Encourage the establishment Of farmer cooperatives and establish links between them and processors, retailers, and Markets. Improving market accessibility guarantees reasonable prices and stability in the Market, which incentivizes more farmers to start millet farming.

8.9.6 Value Addition

To make millet products more appealing to consumers, encourage Value addition. In addition to creating a range of millet-based snacks, drinks, and baked Goods, this can also involve processing millets into flour, flakes, or other handy forms.

8.9.7 Certification and Quality Standards

Establish quality standards and certification Schemes for millet-based products. This increases the market acceptability of millet-Based products by reassuring consumers about their nutritious worth and quality. Additionally, certification can increase export prospects by providing access to foreign Markets.

8.9.8 Financial Support Mechanisms

Create financial assistance systems, such as grants, low-interest loans, and insurance Plans, specifically for millet producers. These systems can serve as safety nets in the event Of unanticipated difficulties, including crop failure brought on by unfavorable weather.

8.9.9 Policy Framework

Create a framework for policy that is helpful and acknowledges the Role that millets play in sustainable agriculture. Pricing, procurement, and land use policies Ought to encourage the growing of millet. Think about implementing laws that give millet Growers a particular share of agricultural subsidies.

8.9.10 Climate-Smart Agriculture Techniques

Stress the use of climate-smart techniques for Growing millet. Promote the application of conservation agriculture, agroforestry, and Organic agricultural techniques to improve soil health, water retention, and overall Resilience to the effects of climate change.

8.9.11 Public Education Initiatives

Start extensive public awareness initiatives to highlight millets as a sustainable and Healthful food choice. Spread the word through a variety of media outlets about the health Advantages and environmentally friendly qualities of millets.

8.9.12 Development and Training

Provide farmers who cultivate millet with training Opportunities and programs aimed at improving their skills. In order to enable farmers to Actively engage in value chains, this involves teaching knowledge on post-harvest Management, processing processes, and entrepreneurial skills.

8.9.13 Integration into Government Programs

Include millets in currently running Government initiatives for nutrition, food security, and poverty reduction. This can entail Adding millet-based goods to public distribution networks, school lunch programs, and Other charity endeavors.

8.9.14 International Collaboration

Encourage cross-border cooperation by taking part in Joint initiatives, research alliances, and knowledge-sharing platforms. Gaining knowledge From other locations' effective millet promotion tactics will hasten the adoption of best Practices and provide insightful information.

8.9.15 Monitoring and Assessment

To follow the development of millet promotion programs, Create strong monitoring and assessment systems. Regular evaluations can assist in Identifying problems, gauging the effectiveness of interventions, and modifying plans of Action as necessary.

8.10 Upcoming Events in Millet Agriculture

Based on general patterns, the following are some prospective future initiatives in millet Agriculture:

8.10.1 Research and Innovation

New and improved millet varieties with increased yield, Nutritional content, and resistance to pests and diseases may be introduced as a result of Ongoing research and innovation in millet cultivation.

8.10.2 Technology Adoption

The millet farming community may continue to embrace more contemporary agricultural Technologies, such as digital platforms for farm management, IoT-based monitoring, and Precision farming.

8.10.3 Sustainable Practice

In order to maintain environmental resilience, expect a greater focus on sustainable Agriculture methods in millet cultivation, such as organic farming, programs all highlight how they can enhance community health. Agro ecology, and water Conservation techniques.

8.10.4 Capacity Building

To empower farmers and increase overall productivity, training sessions and workshops in Millet growing methods, post-harvest management, and value addition may be arranged.

8.10.5 Climate Resilient Agriculture

Climate-resilient millet varieties and farming practices may become more popular in light Of growing worries about climate change. This will help to assure consistent production Under changing weather conditions.

8.10.6 Consumer Awareness Campaigns

Anticipate ongoing initiatives to educate consumers about the health advantages of Millets, possibly in the form of campaigns endorsing millet-based meals, diets, and their Ability to combat malnutrition.

8.10.7 Value Chain Development

To build a more resilient and profitable industry, steps to improve the millet value chain as A whole—from cultivation to processing and marketing—might be taken.

8.10.8 Digital Platform for Farmers

It's possible that more mobile applications and digital platforms catered to millet farmers Will be developed and used, offering them access to weather forecasts, market trends, andreal-time information.

8.10.9 Policy Advocacy for Millet Inclusion

Millets may continue to be pushed for inclusion in nutrition, food security, and sustainable Agricultural policies by groups who support a more resilient and varied food system.

8.11 Conclusion

In conclusion, there is great potential for promoting environmental sustainability, Economic resilience, and enhanced well-being for agricultural communities through the Pursuit of sustainable livelihoods through millet agriculture. Because of their natural Resilience and nutritional advantages, millets are an important part of the agricultural Revolution that aims to achieve sustainable practices. First, millet farming greatly enhances The economic aspect of sustainable lives. Farmers can become less dependent on Monoculture and take advantage of a wider range of market options by expanding their Crop diversification to include millets. The potential economic benefits of millet production Are increased by the increasing demand in healthy, alternative grains around the world. For Farmers to be economically empowered, governments and other agricultural stakeholders Must keep offering financial incentives, low-interest loans, and supportive regulations. The Creation of equitable market connections, farmer cooperatives, and value-adding Programs all support the millet industry's ability to remain economically viable. A vital component of sustainable livelihoods is environmental sustainability, and millet Agriculture fits in nicely with this requirement. As crops with a natural ability to adapt to a Variety of agro-ecological zones, millets are particularly well-suited to areas with limited Water resources and unpredictable weather patterns. Millets are a practical answer to climate change as agriculture struggles with its effects. Incorporating agro ecological methodologies, organic farming techniques, and sustainable Water management strategies into millet agriculture promotes soil health, biodiversity Preservation, and ecological equilibrium. Furthermore, millets' high nutritional content makes them a key component in improving Communal well-being. Encouraging millets as a staple grain can have a major role in Alleviating health issues connected to diet and malnutrition that are plaguing nations. Millets' varied nutritional profile, which includes high levels of fiber, protein, and Micronutrients, is in line with the global trend toward healthier eating habits. Campaigns to Raise public awareness, dietary instruction, and the inclusion of millets in school lunch programs all highlight how they can enhance community health. Another essential component of sustainable livelihoods is community resilience, which is Largely facilitated by millet agriculture. Millet production becomes a community endeavor Through the promotion of knowledge-sharing, farmer cooperative support, and community Engagement. Communities that grow millet are strong enough to withstand economic, Climatic, or social adversities because of their shared responsibility and sense of Empowerment.

When using millet agriculture to navigate the path towards sustainable living, it is critical to Make prudent use of technical improvements. Using contemporary agricultural Technologies can improve productivity, yield, and market accessibility. Examples of these Technologies include digital platforms for information sharing and precision farming. But While integrating technology, it's important to keep inclusion and local contexts in mind, Especially for smallholder farmers.

In conclusion, millet cultivation offers sustainable livelihoods that harmoniously combine Community resilience, environmental stewardship, economic viability, and nutritional well-Being. To push millet cultivation to the forefront of sustainable farming methods, Communities, governments, and agricultural organizations must work together. In addition To securing farming communities' livelihoods, embracing the many benefits of millets helps Ensure a more resilient, equitable, and sustainable future for global agriculture.

References

Amadou, I., Le, G.-W., & Shi, Y.-H. (2013). Millet grain phenolics and their role in disease risk reduction and health promotion. *Food Chemistry, 141*(3), 1919–1927. https://doi.org/10.1016/j.foodchem.2013.05.090

BMI, Fitch Solutions. (2024, March 1). Promotion of minor grains set to increase as food insecurity, climate change concerns mount. *Fitch Solutions: Agribusiness*. Retrieved July 11, 2025.

Cheung, M. M., Deutsch, J., Miller, L., Sherman, R., Katz, S. H., & Wise, P. M. (2025, April 15). *Pearl millet emerges as a healthy, sustainable, gluten free alternative to wheat in the US, according to taste research. Bioengineer*. https://bioengineer.org/pearl-millet-emerges-as-a-healthy-sustainable-gluten-free-alternative-to-wheat-in-the-us-according-to-taste-research

Datawheel. (n.d.). Cereals: millet, other than seed (HS 100820). The Observatory of Economic Complexity. Retrieved July 11, 2025.

Food and Agriculture Organization of the United Nations. (n.d.). One Country One Priority Product: India. FAO Regional Knowledge Platform. Retrieved July 11, 2025, from https://www.fao.org/one-country-one-priority-product/asia-pacific/india/en

Food and Agriculture Organization of the United Nations. (2023). International Year of Millets 2023. FAO. Retrieved July 11, 2025

Food and Agriculture Organization of the United Nations. (2023). Millet value chain development in South Asia: Opportunities and constraints (Open Knowledge Repository Publication No. 9b104e52-de12-4a8e-919b-c4bde32b3cad). FAO

https://www.mymillets.com/Millets-Grains-Flours/Foxtail-Millet-Gluten-Free-Unpolished?srsltid

https://www.ncbi.nlm.nih.gov/pmc/articles/PMC5526919/

https://nutricereals.dac.gov.in/

Nithiyanantham, S., Kalaiselvi, P., Mahomoodally, M. F., Zengin, G., Abirami, A., & Srinivasan, G. (2019). Nutritional and functional roles of millets—a review. *Journal of Food Biochemistry, 43*(e12859). https://doi.org/10.1111/jfbc.12859

Pokharia, A. K., Kharakwal, J. S., & Srivastava, A. (2014). Archaeobotanical evidence of millets in the Indian subcontinent with some observations on their role in the Indus civilization. Journal of Archaeological Science, 42, 442–455. https://doi.org/10.1016/j.jas.2013.11.029 reddit.com

Press Trust of India. (2023, March 18). Millets can help tackle challenges of food security: PM Modi. The Times of India. Retrieved July 11, 2025, from https://timesofindia.indiatimes.com/india/millets-can-help-tackle-challenges-of-food-security-pm-modi/articleshow/98750958.cms

Press Information Bureau. (2023, February 2). APEDA organizes Virtual-Buyer Seller Meet to harness export opportunities in UAE for millets [Press release]. Government of India. https://pib.gov.in/PressReleseDetailm.aspx?PRID=1895783

Sahoo, J. P., & Mahapatra, M. (2023). International Year of Millets – 2023: Revitalisation of millets towards a sustainable nutritional security. Technology in Agronomy, 3, 10. https://doi.org/10.48130/TIA-2023-0010

Sheahan, C. M. (2014). *Plant guide for proso millet (Panicum miliaceum)*. USDA–Natural Resources Conservation Service, Cape May Plant Materials Center. https://plants.usda.gov/DocumentLibrary/plantguide/pdf/pg_pami2.pdf

Singh, B., Sayam, M., Kajla, V., Kashyap, S., Bilandi, A., & Ahmad, M. S. (2024). *The role of millets in diabetes management: From bioactive compounds to metabolic health. Journal of Pharmaceutical Technology, Research and Management, 12*(2), 66–77. https://www.researchgate.net/publication/392731326_The_Role_of_Millets_in_Diabetes_Management_From_Bioactive_Compounds_to_Metabolic_Health

The Biosaline Agronomy article has author(s) & year; if not, you'll need to add those when available.) Author(s). (2023). Title of the article. Agronomy, 15, 165. http://www.biosaline.org/sites/default/files/publicationsfile/agronomy-15-00165.pdf

The Role of Millets in Diabetes Management: From Bioactive Compounds to Metabolic Health. (2023). *ResearchGate*. https://www.researchgate.net/publication/392731326_The_Role_of_Millets_in_Diabetes_Management_From_Bioactive_Compounds_to_Metabolic_Health

United Nations Zambia. (2023, July 10). Commemoration of the International Year of Millets 2023. United Nations.

Yadav, O. P., Singh, D. V., Kumari, V., Prasad, M., Seni, S., Singh, R. K., Sood, S., Kant, L., Rao, B. D., Madhusudhana, R., Bhat, B. V., Gupta, S. K., Yadava, D. K., & Mohapatra, T. (2024). Production and cultivation dynamics of millets in India. *Crop Science*, 1–26. https://doi.org/10.1002/csc2.21207

Zhang, A., Ji, Y., Sun, M., Lin, C., Zhou, P., Ren, J., Luo, D., Wang, X., Ma, C., Zhang, X., Feng, G., Nie, G., & Huang, L. (2021). Research on the drought tolerance mechanism of *Pennisetum glaucum* (L.) in the root during the seedling stage. *BMC Genomics, 22*, Article 568. https://doi.org/10.1186/s12864-021-07888-5

Millet Marvel: Unveiling the Superfood's Potential

9

Shubhangi Nigam, Nandini Saini, and Khushboo Mansoori

Abstract

This paper explores the nutritional and superfood potential of millets, investigating their macronutrient and micronutrient composition. Millets stand out as versatile grains, abundant in complex carbohydrates, proteins, and vital vitamins and minerals. This nutritional richness positions millets as a holistic dietary choice, offering sustained energy and supporting overall well-being. A comparative analysis with other grains and cereals highlights the nutritional superiority of millets, showcasing lower glycemic index values and enhanced micronutrient profiles.

Noteworthy is the gluten-free and hypoallergenic nature of millets, making them inclusive for individuals with dietary restrictions. Furthermore, their environmental sustainability enhances their significance as a resilient agricultural option. The paper concludes by emphasizing the transformative impact millets can have on global nutrition, advocating for increased awareness and accessibility. It envisions a future where millets play a central role in fostering a healthier and more sustainable food landscape. To unlock their full potential, the paper underscores the need for ongoing research, culinary innovation, and collaborative efforts. As we unravel the marvel of millets, this research contributes to the growing discourse on ancient grains, envisioning a shift towards a more nourishing and sustainable global food system.

S. Nigam (✉) · N. Saini · K. Mansoori
Department of Food Technology Institute of Engineering and Technology, Bundelkhand University, Jhansi, UP, India

C. B. Singh et al. (eds.), *Global Millets Production for a Sustainable Future*,
https://doi.org/10.1007/978-3-031-88157-2_9

Keywords

Nutritional potential · Dietary inclusivity · Low glycemic index · Resilient agriculture · Culinary innovation · Health and wellbein

9.1 Introduction

This paper explores the nutritional and superfood potential of millets, analyzing their macronutrient and micronutrient composition. Millets are versatile grains rich in complex carbohydrates, proteins, and essential vitamins and minerals, making them a holistic dietary choice that supports sustained energy and overall well-being. A comparative analysis with other grains and cereals underscores their nutritional advantages, including a lower glycemic index and superior micronutrient profile.

Additionally, the gluten-free and hypoallergenic nature of millets makes them suitable for individuals with dietary restrictions. Their environmental sustainability further highlights their importance as a resilient and eco-friendly agricultural crop. The paper concludes by emphasizing the transformative potential of millets in addressing global nutrition challenges, advocating for increased awareness, accessibility, and adoption.

9.2 Nutritional Composition of Millets (Table 9.1)

9.2.1 Macronutrients

9.2.1.1 Carbohydrates

Millets are rich in complex carbohydrates, providing sustained energy. They have a lower glycemic index compared to some other grains, which can be beneficial for blood sugar control. This characteristic makes millets an excellent choice for those seeking to manage diabetes or maintain stable energy levels throughout the day (Ariane Lang, 2023).

9.2.1.2 Proteins

Millets contain a good amount of protein, making them a valuable source for vegetarians. The protein content varies among different millet varieties but is generally comparable to or higher than other cereals. This makes millets a viable alternative for those looking to increase their protein intake without consuming animal products (Hassan et al., 2021).

9.2.1.3 Fats

Millets are low in fat, with most of the fat being unsaturated. This can contribute to heart health when included as part of a balanced diet. The low-fat content, combined

Table 9.1 Macronutrient and energy content of millets

Millets	Moisture (g)	Protein (N*6.25)	Fat (g)	Mineral (g)	Dietary Fibre (g)	Carbohydrate (g)	Energy (kcal)
Finger millet	13.1	7.3	1.3	2.7	11.5	72.0	328
Proso millet	11.9	12.5	1.1	1.9	–	70.4	341
Foxtail millet	11.2	12.3	4.3	3.3	2.4	60.9	331
Little millet	11.5	7.7	4.7	1.5	2.53	67.0	341
Kodo millet	12.8	8.3	1.4	2.6	2.47	65.9	309
Barnyard millet	11.9	6.2	2.2	4.4	1.98	65.5	307
Brown-top millet	11.3	9.0	1.9	3.9	–	71.3	338

Source: https://www.researchgate.net/publication/364401665/figure/tbl1/AS:1143128109102569 9@1666274223998/Macronutrient-contents-of-millets-with-major-cereals-rice-and-wheat.png

with a high level of unsaturated fats, makes millets an ideal choice for those aiming to maintain a healthy lipid profile (Brahmanad Nayak, 2023).

9.2.2 Micronutrients

9.2.2.1 Vitamins

Millets are a good source of various vitamins, including B vitamins such as niacin, thiamine, and riboflavin. These vitamins play crucial roles in energy metabolism and overall health. B vitamins are essential for converting food into energy, supporting nervous system health, and maintaining healthy skin and eyes (Millets Heal, 2023).

9.2.2.2 Minerals

Millets contain essential minerals like iron, magnesium, phosphorus, and zinc. These minerals are important for bone health, immune function, and overall well-being. Iron supports red blood cell production, magnesium aids muscle function, phosphorus is vital for bone strength, and zinc enhances immune response.

9.3 Comparison with Other Grains and Cereals

9.3.1 Millets vs. Wheat (essential.com 2023)

- **Nutrient Content:** Millets generally have higher levels of essential nutrients such as iron and magnesium compared to wheat. These minerals are crucial for functions like oxygen transport, muscle function, and maintaining energy levels.

- **Gluten-Free Alternative:** Unlike wheat, millets are naturally gluten-free, making them a suitable option for individuals with celiac disease or gluten sensitivity. This aspect significantly broadens their dietary inclusivity.

9.3.2 Millets vs. Rice (essential.com 2023)

- **Glycemic Index:** Millets often have a lower glycemic index than rice, which helps in managing blood sugar levels more effectively. This makes millets a better choice for individuals with diabetes or those looking to maintain stable energy levels.
- **Micronutrient Density:** While rice is a staple in many diets, millets offer a richer array of nutrients, including fiber and essential vitamins and minerals, which contribute to overall health.

9.3.3 Millets vs. Quinoa (essential.com 2023)

- **Nutrient Diversity:** Both millets and quinoa are excellent sources of protein and fiber, beneficial for digestion and muscle repair. However, millets tend to have a more diverse nutrient profile, offering a wider range of vitamins and minerals.
- **Cost-Effectiveness:** Millets are often more affordable than quinoa, making them an accessible option for many people seeking nutritious food options without high costs.

9.3.4 Millets vs. Oats (essential.com 2023)

- **Fiber and Nutrient Range:** While oats are renowned for their high fiber content, which aids digestion and heart health, millets offer a more comprehensive range of nutrients. Millets provide a broader spectrum of vitamins and minerals, making them an excellent addition to a varied diet.
- **Gluten-Free Benefits:** Like oats, millets are naturally gluten-free, offering a safe and nutritious alternative for those avoiding gluten (Fig. 9.1).

9.4 Health Benefits

9.4.1 Millets and Diabetes Management Diabetes

India has the second-highest prevalence of diabetes, only after China, with approximately 77 million Indians affected by it. Lifestyle modifications with dietary intervention are key in managing Type 2 diabetes. The high fiber content in millets aids in the slow release of sugars into the bloodstream, maintaining a low postprandial glucose level. The B vitamins in millets support better carbohydrate utilization, and

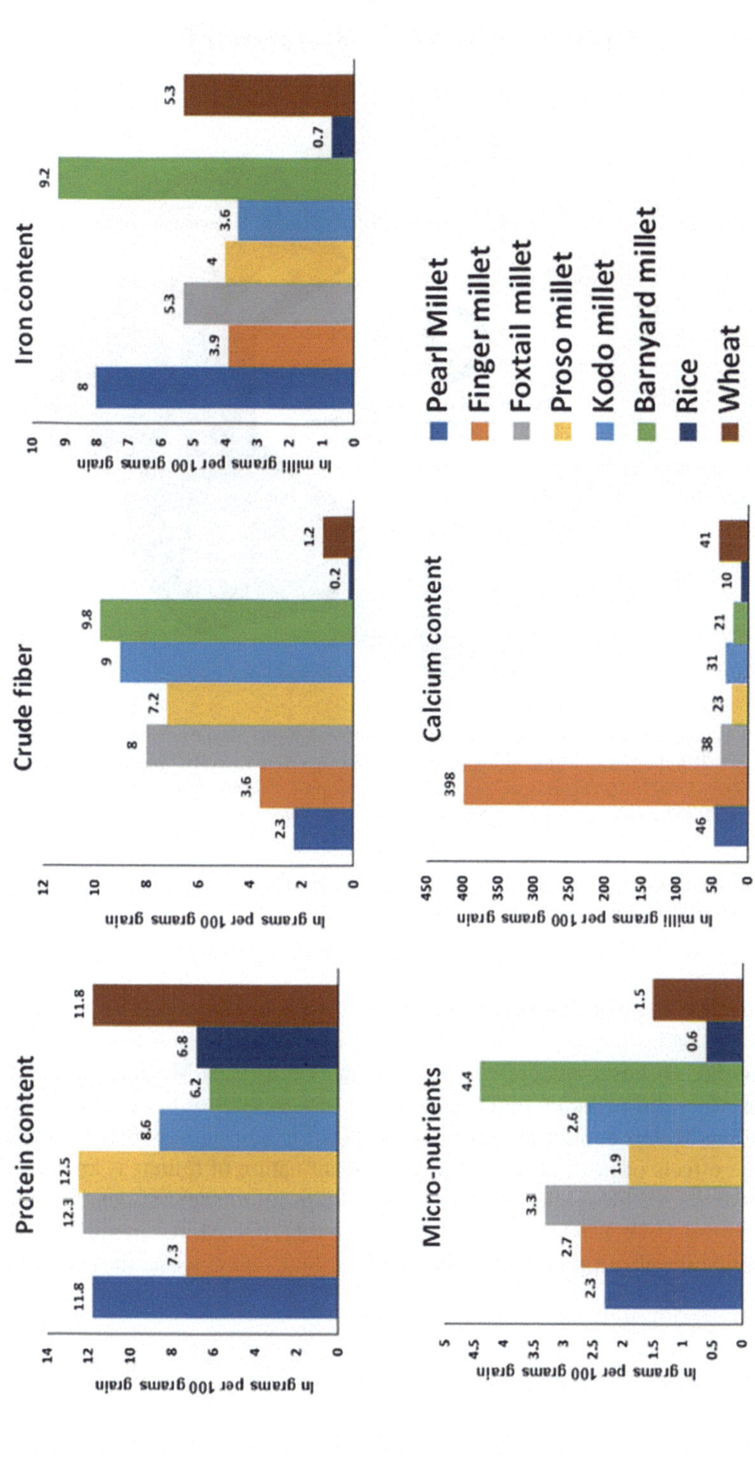

Fig. 9.1 Comparative illustration of the nutritional content of millet seeds (per 100 g grain weight) with rice and wheat. Source: https://www.researchgate.net/publication/349622498/figure/fig1/AS:11431281153093317@1682323536959/Fig-1-Comparative-illustration-of-the-nutritional-content-of-millet-seeds-per-100-g.jpg

Fig. 9.2 Millets health benefits. Source: https://www.google.com/url?sa=i&url=https%3A%2F%2Fapeda.gov.in%2Fmilletportal%2Fhealth_benefits.html&psig=AOvVaw2RGbbEAqAYfJsFdErupKht&ust=1726848859703000&source=images&cd=vfe&opi=89978449&ved=0CBQQjRxqFwoTCLjCubizz4gDFQAAAAAdAAAAABAE

ultimately millets serve as a low glycemic food, especially effective for Type 2 diabetes (Dansinger & Watson, 2004) (Fig. 9.2).

9.4.2 Cardiovascular Benefits of Millets

Cardiovascular Disease: According to the Indian Council of Medical Research and Registrar General of India, India accounts for almost 60% of the world's heart disease burden. Millets contain significant amounts of magnesium, which is known to reduce the effects of heart attacks. The antioxidant nature of millets is known to help reduce oxidative stress, and the fiber in them helps minimize the accumulation of bad cholesterol (LDL) and increase good cholesterol (HDL). Fiber in millets also aids in delaying the absorption of cholesterol, dietary fat, and glucose, thus contributing to heart health (Tanushree Banerjee, 2022).

9.4.3 Hypertension Control with Millets

Hypertension: According to the National Family Health Survey-5, 24% of men and 21% of women over the age of 15 years in India are hypertensive. High blood pressure is a risk factor for heart diseases and stroke. The DASH diet (Dietary Approaches to Stop Hypertension) promotes the consumption of high-fiber whole foods. Research suggests the benefits of millet consumption in controlling hypertension. Millets are rich in dietary fiber and contain certain antioxidants and micronutrients like magnesium and calcium that are effective in maintaining normal blood pressure. They also ensure a good serum lipid profile (Mohammad & Bansod, 2024).

9.4.4 Millets for Celiac Disease

Celiac Disease: Gluten intolerance is seen in those with gastrointestinal disorders like Celiac disease. Gliadin in wheat, hordein in barley, secalin in rye, and avenin in oats are from the family of gluten proteins that cause inflammation of the inner lining of the intestine in gluten-intolerant people. Being naturally gluten-free, millets serve as an excellent replacement for cereal grains in those suffering from Celiac disease (Balakireva & Zamyatnin, 2016) (Fig. 9.3).

9.5 Antioxidant and Anti-inflammatory Properties

9.5.1 Millets and Metabolic Health: The Antioxidant Advantage

Millet-based diets are rich in antioxidants, which offer potential for reducing inflammation and providing protection against chronic ailments such as heart disease and cancer. The spectrum of benefits from millet derived foods includes their anti-hypertensive, anti-inflammatory, antimicrobial, hypo cholesterolemic, hypoglycemic, andante—oncogenic properties, coupled with their capacity to positively impact gut health through immune modulation. A systemic review suggests that, millet consumption over a period of 21 days to 4 months, there is a significant reduction in total cholesterol (TC), triacylglycerol, high-density lipoprotein cholesterol (HDL-C), low-density lipoprotein cholesterol (LDL-C), and very-low–density lipoprotein cholesterol (VLDL-C).In four investigation t is also observed that millets ingestion has normalized TC and triacyl glycerol levels (<200 and <150 mg/dl, respectively). Millet-based meals also increased HDL-C by 6.0%, lowered blood pressure by 4.0 and 5.0%, and lowered BMI by 7.0%.Overall, these studies suggest that integrating millet-based items into the diet has the potential to support in the control of metabolic disease. It is crucial to emphasize, however, that millet should be consumed in moderation as part of a well-balanced diet, and persons with metabolic disorders should consult a healthcare provider before making dietary adjustments (Front. Nutr., 2021) (Figs. 9.4 and 9.5).

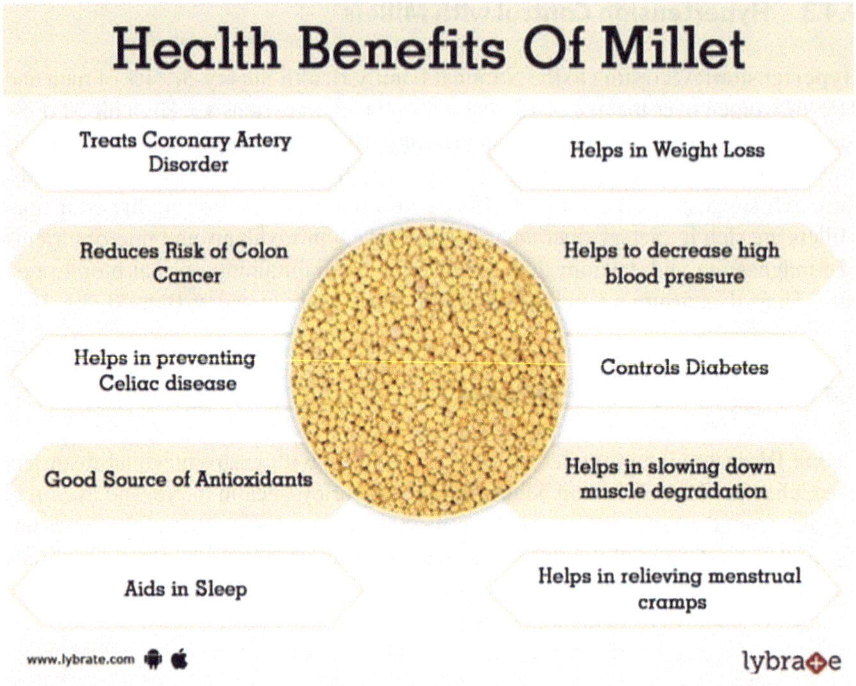

Fig. 9.3 Health benefits of millet. Source: https://wp-public-fs.s3.ap-south-1.amazonaws.com/tasks/15f5f3fc928d150e4ca2f733dc707651effa2562c4c846/images/image-5f60a5ae07c93.png

9.6 Gluten Free and Hypoallergic Nature

9.6.1 The Challenges of Gluten-Free Alternatives

However, the structural, functional, and rheological properties offered by gluten to the dough could not be replaced. Hence, this study suggests the use of millets as an alternative diet. Millets not only protect against gluten sensitivities but also provide all other essential micro- and macronutrients required for the growth and metabolism of the body. They hold potential in eradicating malnutrition due to their high nutritional, phenolic, and other antioxidant elements, indicating their probability to help in achieving a major sustainable goal related to good health and well-being (Abang Zaidel et al., 2010).

9.6.2 The Potential of Millets

Despite offering a wide variety of health benefits, the consumer acceptability of millets is quite low at the moment. This paper is a wake-up call for researchers and farmers to gear up in popularizing millets and their associated products and in

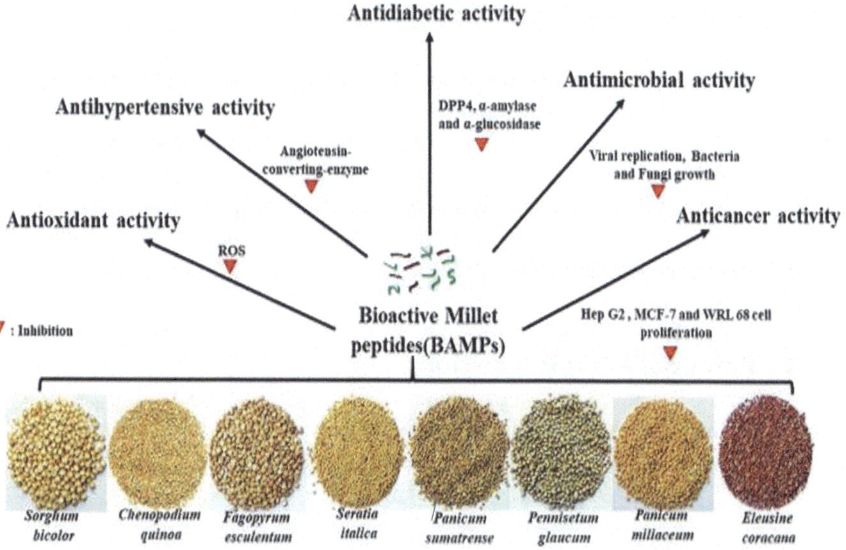

Fig. 9.4 Functional properties and health benefits. Source: https://www.researchgate.net/publication/337057492/figure/fig1/AS:824689626718209@1573632737581/Health-promoting-biological-activities-of-bioactive-millet-peptides-BAMPs.jpg

Fig. 9.5 Antioxidant and anti-inflammatory properties of millets. Source: https://buzzle.com/

improving the limitations associated with the shelf life of certain millets. Different physical processing technologies in the production of traditional and modern millet-based food products are also discussed in this paper, providing an overview of how various major millets can be used to promote health security with lesser dependency on environmental conditions for their growth and production (Siroha & Bangar, 2024).

9.6.3 Nutritional and Health Benefits of Millets

Millets are a group of small-seeded grains that are gluten-free and hypoallergenic. They are rich in nutrients like vitamins, minerals, dietary fiber, and antioxidants. Millets are a good source of dietary protein, with a balanced amino acid profile. They are considered a safe option for people with allergies, especially those related to gluten or wheat. They are also suitable for people with diabetes or gluten intolerance because they have a low glycemic index. Millets are easy to digest and soothing on the stomach (Ariane Lang, 2022).

9.6.4 Potential Allergic Reactions

However, some cases of anaphylaxis due to millet consumption have been reported. Cross-reactivity between millet and other cereals is also not uncommon (Fig. 9.6).

9.7 Culinary Versatility

9.7.1 Types of Millets

9.7.1.1 Finger Millet (Ragi)

Rich in calcium, iron, and fiber, finger millet, or ragi, is a winter-friendly millet that provides essential nutrients. The high fiber content aids digestion, preventing digestive issues commonly associated with cold weather. Ragi is also a good source of warmth, making it an ideal choice for the winter season. It can be consumed in the form of porridge, rotis, or added to soups for a nutritious boost (Times of India, 2024).

9.7.1.2 Pearl Millet (Bajra)

Pearl millet, commonly known as bajra, is a staple winter grain in many regions. It is rich in iron, magnesium, and phosphorus, essential minerals that contribute to overall health. Bajra is known for its warming properties, making it an excellent choice for winter consumption. It can be used to prepare rotis, khichdi, or even added to soups and stews for a hearty and nutritious meal.

9.7.1.3 Sorghum (Jowar)

Jowar, or sorghum, is a gluten-free millet that is highly versatile and nutritionally dense. It is an excellent source of dietary fiber, protein, and antioxidants. Jowar's complex carbohydrates provide sustained energy, keeping the body warm during colder days. Including jowar in the form of rotis, porridge, or upma can be a nutritious way to combat winter lethargy and boost overall health (Times of India, 2024) (Fig. 9.7).

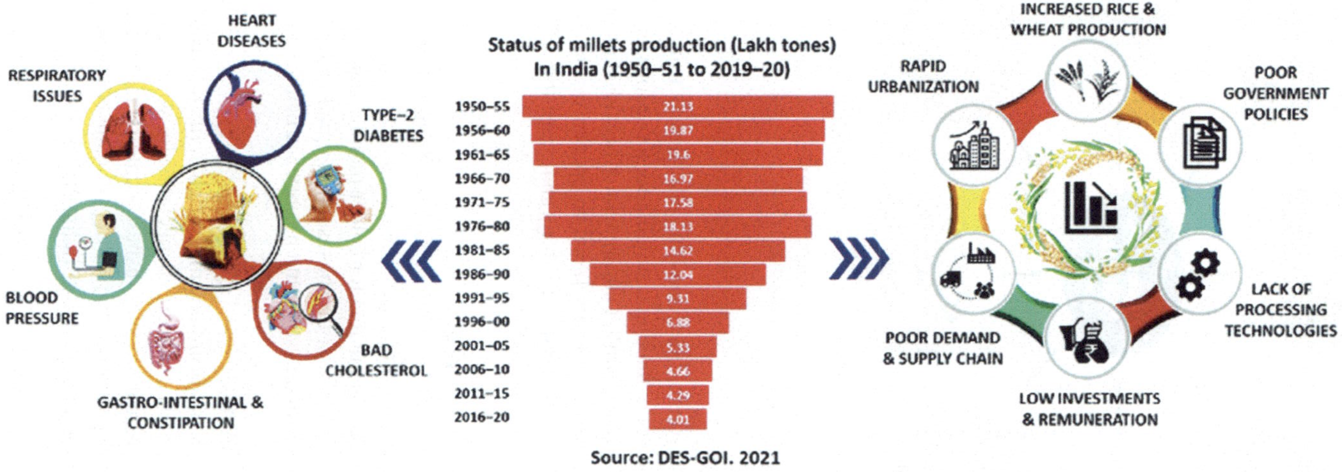

Fig. 9.6 Millets: Health benefits, production, and challenges in India. Source: https://www.researchgate.net/profile/Nanje-Na-2/publication/358500139/figure/fig1/AS:1122661988085771@1644674884811/Millets-health-benefits-production-and-challenges-in-India-Data-taken-from-various.png

Fig. 9.7 Types and benefits of millets. Source: https://www.pinterest.com/pin/types-and-benefits-of-millets-nutritionfactin%2D%2D624171448280573771/

9.7.1.4 Foxtail Millet

Foxtail millet is rich in protein, fiber, and minerals such as iron and magnesium. It is a low-glycemic-index grain, making it suitable for those looking to manage blood sugar levels. Consuming foxtail millet during winter provides a sustained release of energy, helping to keep the body warm. It can be used in various recipes, including pulao, khichdi, and dosa, offering a healthy and warming alternative to traditional grains.

9.7.1.5 Proso Millet (PaniVaragu)

Proso millet, also known as panivaragu, is a nutritious millet option for the winter season. It is a good source of vitamins, minerals, and antioxidants. The warming nature of proso millet makes it suitable for colder months, and its versatility allows it to be incorporated into a variety of dishes, such as upma, pulao, or even desserts. The high fiber content aids in digestion and promotes a healthy gut.

9.7.1.6 Little Millet (Samai)

Little millet, or samai, is a small-grained millet rich in fiber, B-vitamins, and minerals. Its nutritional profile makes it an excellent choice for winter consumption. Little millet can be used to prepare a range of dishes, including rice replacements, upma, and porridge. Its warming properties and nutrient content contribute to overall well-being during the colder months (Times of India, 2024).

9.7.1.7 Kodo Millet

Kodo millet is a nutrient-dense millet that provides a good amount of protein, fiber, and essential minerals. It is known for its ability to regulate blood sugar levels, making it a suitable option for individuals with diabetes. Including kodo millet in winter meals can offer sustained energy and warmth. It can be used to prepare dishes like pulao, khichdi, or as a rice substitute in various recipes (Times of India, 2024).

9.7.1.8 Barnyard Millet (Sanwa)

Barnyard millet, also known as sanwa, is a gluten-free grain rich in fiber, iron, and calcium. It is easy to digest and provides a quick energy boost. The warming nature of barnyard millet makes it an ideal choice for winter. It can be used in the preparation of upma, khichdi, or even desserts, offering a nutritious alternative to traditional grains (Barnyard Millet, 2017).

9.7.2 Incorporating Millets Into Your Winter Diet

9.7.2.1 Porridge

Millet porridge made with any of the mentioned millets can be a comforting and nutritious breakfast option. Sweeten it with honey or add fruits for added flavor and health benefits.

9.7.2.2 Rotis and Bread

Millet flour can be used to make rotis or bread, providing a wholesome alternative to refined wheat flour. This is an excellent way to include millets in daily meals.

9.7.2.3 Khichdi

Prepare a nutritious and warming khichdi using a combination of millets, lentils, and vegetables. Season it with warming spices like cumin and black pepper.

9.7.2.4 Pulao

Millets can be used as a base for pulao, incorporating a mix of vegetables, herbs, and spices to create a flavorful and hearty dish.

9.7.2.5 Upma

Millets can be used in place of semolina to make a healthy and warming upma. Add vegetables, nuts, and spices for added taste and nutrition (Smitha Kalluraya, 2015).

9.7.2.6 Soups and Stews

Add cooked millets to soups and stews to enhance their nutritional content. The grains absorb the flavors of the broth, creating a wholesome and filling dish.

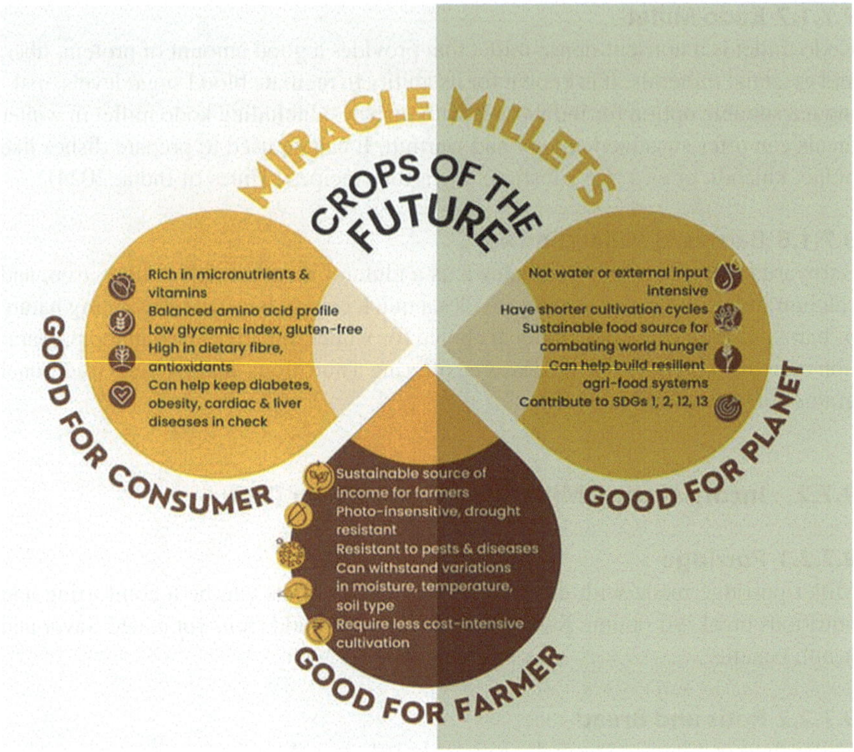

Fig. 9.8 Millets the future of sustainable agriculture. Source: https://www.civilsdaily.com/news/millets/

9.7.2.7 Desserts

Millets can be used in dessert recipes, such as puddings, kheer, or even as a base for healthy energy bars. This allows for a sweet treat while still reaping the nutritional benefits.

9.8 Environmental Sustainbility (Fig. 9.8)

9.9 Environmental Sustainability of Millet Cultivation

9.9.1 Water Efficiency and Environmental Impact

Millets exhibit remarkable water efficiency, thriving in semi-arid conditions with minimal water requirements compared to conventional crops. Their ability to grow in diverse climates contributes to agricultural resilience. Cultivating millets often involves fewer chemical inputs, reducing environmental impact. This inherent adaptability and eco-friendly cultivation make millets a sustainable choice amid growing concerns about water scarcity and environmental sustainability in agriculture (tracex, 2023).

9.9.2 Climate Resilience and Biodiversity Conservation

Millets demonstrate resilience to climate change due to their adaptability to varied environmental conditions. These hardy crops are known for withstanding drought, high temperatures, and poor soil quality. Their minimal water requirements and ability to thrive in challenging climates make millets a crucial resource in ensuring food security and agricultural sustainability in the face of changing climate patterns. Cultivating millets contributes to biodiversity conservation by preserving traditional, locally adapted varieties. These diverse millet strains often have unique genetic traits, enhancing resilience to pests and diseases. By promoting the cultivation and consumption of various millet varieties, communities play a role in safeguarding agricultural biodiversity, ensuring a sustainable and resilient food system for the future (tracex, 2023) (Fig. 9.9).

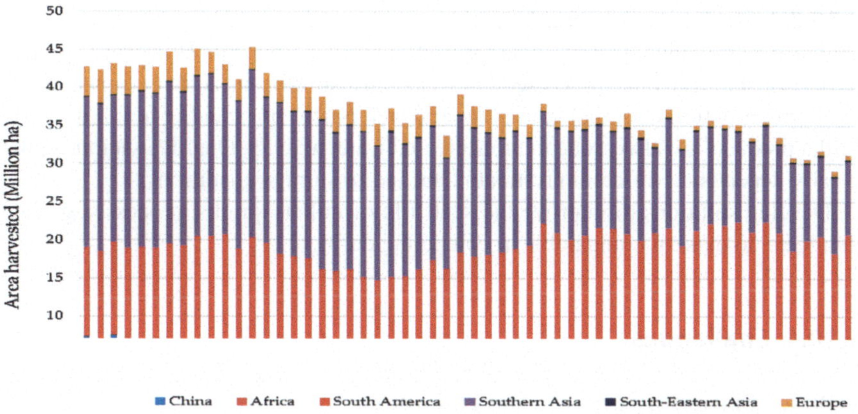

Fig. 9.9 Millets for food security in the context of climate change. Source: https://www.google.com/url?sa=i&url=https%3A%2F%2Fwww.mdpi.com%2F2071-1050%2F10%2F7%2F2228&psig=AOvVaw0Hahq0R879Fgt2DNEoa37P&ust=1726850201382000&source=images&cd=vfe&opi=89978449&ved=0CBcQjhxqFwoTCJDw68O4z4gDFQAAAAAdAAAAABAE

9.10 Challenges and Future Research

9.10.1 Challenge 1

Processing converts the inedible grain into edible form and thereby enhancing its quality. Processing of millets plays significant role during its utilization as food. Longer shelf life, aesthetics of food and flavour and ease of cooking is what necessitates processing.

However some millets require multiple processing for optimization of grain recovery and optimization of polishing to retain their nutrition value Develop solutions to improve the efficiency of millet processing (dehulling, separation, polishing etc.) while retaining the nutritional value.

9.10.2 Challenge 2

9.10.2.1 Decentralized Processing

Processing of millets face several hurdles owing to variation in size of various millet types and low shelf life of the processed millets. The grains vary in terms of shape, nature of grain surface, hardness, husk-grain bonding etc. Furthermore, there are variations within the same small millet crop due to variation in varieties, cultivation practices, and microclimate across production regionsDevelopmachineryand/orinfrastructureforsmallscaledecentralizedprocessingthatisaffordableandeasy to use for farmers, FPOs and consumers.

9.10.3 Challenge 3

9.10.3.1 Shelf Life Augmentation

Milletsareextremelynutritiousandareproventohavehealthbenefits.However, millets have poor shelf life once processed due to its intrinsic enzyme activity (lipase activity, lipid oxidation etc.) that causes rapid development of rancidity and bitterness. Millet products are also prone to moisture and water activity. Quality assurance thus greatly depends on different pre-treatments and/or storage conditions.

Develop solutions to improve the shelf life of millets–grains, processed grains, flours etc. to make it comparable to competing crops.

9.10.4 Challenges

9.10.4.1 Marketing and Market Linkages

Millet supply chain suffers from inconsistent supply and demand that prevents its commercial viability. While the lack of access to HYV seeds has led to low crop productivity, the lack of public awareness about nutritional benefits of millets has led to limited adoption of millets as a ready to cook cereal. In addition, limited

Fig. 9.10 Wonder world of millets. Source: https://www.slideshare.net/slideshow/wonder-world-of-millets/3645386

distribution and lack of market knowledge have resulted in sub-optimal reach, lower price realization and wastage (Fig. 9.10).

9.11 Nutritional Enhancement

9.11.1 Biotechnological and Sustainable Approaches

Biotechnological approaches are being explored to enhance the nutritional value of millet grains, aiming to improve their health benefits and meet evolving dietary needs. Sustainable production practices are also critical, focusing on promoting research and development to ensure resilient and eco-friendly cultivation of millets. This involves improving varieties, extending shelf life, optimizing processing methods, and enhancing market access to support the broader adoption of millet-based products (Sustainable Food Technol., 2024).

9.11.2 Historical and Resilient Traits

Historically, research has centered on increasing millet yields, with significant advancements made in the 1970s and 1980s through the development of hybrids and improved varieties of sorghum, pearl millet, and finger millet. Millets are notably climate-resilient, capable of withstanding drought, diseases, and pests, and requiring minimal care. As C4 plants, they exhibit higher photosynthetic efficiency compared to C3 plants, converting CO2 into carbohydrates more effectively, which contributes to their adaptability and sustainability (RUBISCO, 2017) (Fig. 9.11).

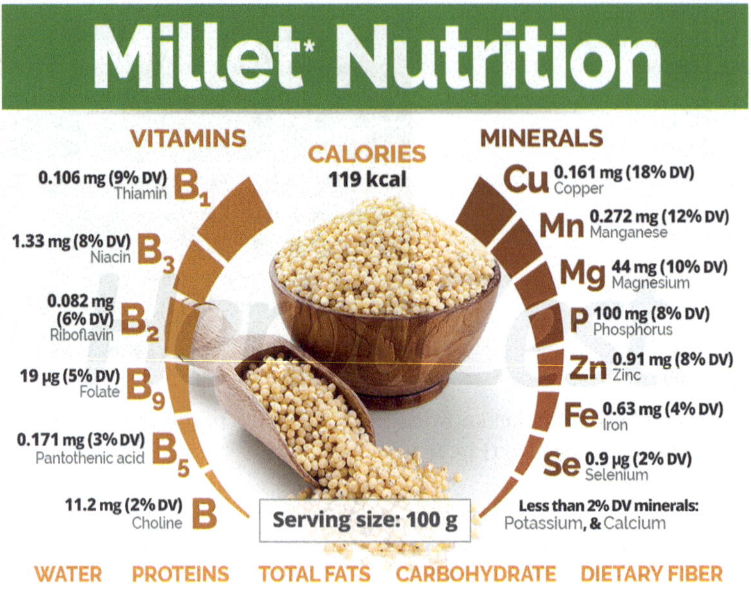

Fig. 9.11 Nutritional enhancement of millets. Source: https://www.pinterest.com/pin/millet%2D%2D417568196715029094

9.12 Conclusion

The exploration of millets as superfoods emphasizes their potential to address global nutritional and environmental challenges. Their rich composition of complex carbohydrates, proteins, essential micronutrients, and bioactive compounds establishes them as a sustainable source of energy and health. Millets' gluten-free nature and low glycemic index further extend their inclusivity to diverse dietary needs.

Millets also demonstrate environmental resilience, thriving in semi-arid conditions with minimal water and input requirements, making them vital for sustainable agriculture and food security. By comparing their nutritional profile with other staple grains, millets stand out for their superior content of iron, magnesium, and dietary fiber, reinforcing their status as a modern-day superfood.

While millets hold immense promise, addressing challenges like limited awareness, processing inefficiencies, and market accessibility is crucial. Collaborative efforts in research, innovation, and policy can unlock their potential, fostering a shift towards healthier diets and sustainable practices.

Ultimately, embracing millets can improve individual health, enhance food security, and contribute to the global fight against climate change, positioning these ancient grains as a cornerstone of future agricultural and nutritional strategies!

References

Journal Articles & Research Papers

Gowda, N. N., Siliveru, K., & Prasad, P. V. (2021). Modern processing of Indian millets: A perspective on changes in nutritional properties. *Food, 11*(4), 499. https://doi.org/10.3390/foods11040499

Hassan, Z. M., Sebola, N. A., & Mabelebele, M. (2021). The nutritional use of millet grain for food and feed: A review. *Agriculture & Food Security, 10*(1), 16. https://doi.org/10.1186/s40066-021-00289-8

Sharma, P., Gujral, H. S., & Kumar, A. (2015). Influence of hydrothermal treatment on starch, dietary fiber, and mineral profile of barnyard millet. *Journal of Food Science and Technology, 52*(7), 4467–4473. https://doi.org/10.1007/s13197-014-1520-8

Saleh, A. S., Zhang, Q., Shen, Q., & Liu, Y. (2013). Millet grains: Nutritional quality, processing, and potential health benefits. *Comprehensive Reviews in Food Science and Food Safety, 12*(3), 281–295. https://doi.org/10.1111/1541-4337.12012

Government & Institutional Websites

International Crops Research Institute for the Semi-Arid Tropics (ICRISAT). (n.d.). Millets for sustainable agriculture. *ICRISAT*. https://www.icrisat.org/

Indian Institute of Millets Research (IIMR). (2023). The role of millets in global food security. *IIMR*. https://millets.res.in/.

Food and Agriculture Organization (FAO). (2023). Millets and food security. *FAO Reports*. https://www.fao.org/.

News & Industry Reports

Times of India. (2024). How India is reviving millet cultivation for better nutrition. *The Times of India*. https://timesofindia.indiatimes.com/.

Better with Millets. (2023). The rise of millets in modern diets. *Better With Millets*. https://www.betterwithmillets.com/.

Tracex Technologies. (2023). Digital solutions for millet supply chains. *Tracex*. https://tracextech.com/.

Other Links

https://www.webmd.com/diabetes/millet-diabetes-blood-sugar?

https://economictimes.indiatimes.com/magazines/panache/cardiovascular-diseases-killing-indians-the-most-pulmonary-embolism-rheumatic-condition-and-other-heart-disorders-people-should-know-about/articleshow/94509483.cms?

https://bmcpublichealth.biomedcentral.com/articles/10.1186/s12889-024-20097-5?

https://www.mdpi.com/2072-6643/8/10/644?

https://tracextech.com/about-us/

https://pubs.rsc.org/en/content/articlelanding/2024/fb/d4fb00047a?

https://www.archanaskitchen.com/6-recipes-to-cook-from-healthy-barnyard-millet?

https://smithakalluraya.com/little-millet-upma-samai-upma-millet-recipes/?

https://www.healthline.com/nutrition/what-is-millet?

https://agricultureandfoodsecurity.biomedcentral.com/articles/10.1186/s40066-020-00282-6?

https://essentiaeats.com/blogs/blog/comparing-millet-with-other-grains-nutritional-analysis?

Millets Grains for Promoting Health of India and SDG

10

Dinesh Kumar Gupta

Abstract

Millet is a highly nutritious, and environmentally friendly food item that contributes significantly to achieving the Sustainable Development Goals. Its production cost is also low and it is a highly nutritious food. It is considered a good source of essential vitamins and minerals. Which helps prevent chronic diseases and promote good health. Millets require fewer resources to produce than other crops, and can help reduce greenhouse gas emissions and address climate change. The Government of India and State Governments are promoting the inclusion of millet in people's daily diet so that food systems can be strengthened. Millets will be very helpful in achieving the Sustainable Development Goals. Under which SDG 2-(Freedom from hunger) SDG 3-(Good health and happiness) SDG 12-(Sustainable consumption and production) and SDG 13-(Climate action).

Keywords

Millet · SDG · Good health · Economic growth · Climate resilient

10.1 Introduction

Millet is a highly nutritious, and environmentally friendly food item that contributes significantly to achieving the Sustainable Development Goals. Its production cost is also low and it is a highly nutritious food. Millets (Bajra, Jowar, Ragi etc) are one of the oldest foods known to humans. Millets were among the first crops to be domesticated in India with several pieces of evidence pointing to their consumption during the Indus Valley civilization (Banerjee & Maitra, 2020). Crops with small and coarse seeds like Sawan, Kangni, China, Jowar, Bajra, Ragi, Kodo, Kutki and Kuttu

D. K. Gupta (✉)
Department of Economics, Rajkiya Mahavidyalaya Amori, Champawat, Uttarakhand, India

C. B. Singh et al. (eds.), *Global Millets Production for a Sustainable Future*,
https://doi.org/10.1007/978-3-031-88157-2_10

are called millet crops or coarse grains. At present, millets are grown in the Indian states of Uttar Pradesh, Madhya Pradesh, Chhattisgarh, Maharashtra, Gujarat, Rajasthan, Haryana, Andhra Pradesh, Telangana, Karnataka and Tamil Nadu (Drishti IAS, 2023). Coarse grains are hardy crops even in marginal moisture, low fertiliser and adverse climates. These crops can be easily grown in regions like rain-fed, dry, hilly, or even coastal areas. They can even be produced on dry land with minimal maintenance (Kumar et al., 2018). Compared to other grains, these crops are least affected by pests and major weather changes. Insects attack them sparingly, hence pesticides are used sparingly. Coarse grain seeds can be stored for many years. Due to this, it is beneficial for areas suffering from drought. This crop can be grown easily with less water and minimum fertiliser use. Coarse grains are good sources of many nutrients like energy, carbohydrates, fats, proteins, fibre, antioxidants, iron, zinc and vitamins etc. This can help eliminate vitamin and mineral deficiencies and malnutrition in India (Mounika & Uma Devi, 2019) and other developing countries. Therefore, people around the world should now pay attention to the production of climate-friendly millets and their general benefits, as they are rich in polyunsaturated acids and omega-3 acids. It helps in reducing cholesterol.

10.2 Types of Millet in India

- **Barnyard millet** is primarily a high source of iron and fiber. Barnyard millet, which is also known by the names Jhangora, Udalu, Oodalu, Shyama etc. in India, is rich in nutrients (Saxena et al., 2018). According to Ayurveda, this millet is sweet in taste and easy to digest, which contributes highly to increasing Vata but balances Pitta and Kapha. When cooked, this small millet resembles rice and is often eaten with curries.
- **Finger millet** is a great alternative to oats and grains as a staple. Ragi or millet is a nutritious food item which contains all the essential macro and micro nutrients (Sharma et al., 2021). Besides, it has very low levels of cholesterol and sodium which helps in keeping the heart healthy.
- **Foxtail millet** which is rich in minerals and vitamins. Foxtail millet contains a lot of fiber, which can relieve constipation and regulate digestion. It helps in weight loss. The fiber present in this millet keeps you feeling full throughout the day. This way, it will reduce your appetite and your weight will also reduce (United Nations, 2023).
- **Little millet** is also rich in iron and fiber. Kutki or little millet (Panicum sumatranse), also known as small millet, is a type of millet that is native to India and is known for its high nutritional value and drought tolerance (YOJANA (2023).
- **Proso millet**, also known as proso millet rice, is a unique Indian millet that was domesticated as a crop thousands of years ago. Apart from proso, this millet is also known as broomcorn millet, common millet and hog millet (Fig. 10.1).

Fig. 10.1 Major and minor varieties of millets. Source: Millets: Empowering the society with nutrient-rich superfoods to achieve sustainable development goals

10.3 Millets and SDG

Due to the important contribution of millets in achieving the Sustainable Development Goals (SDGs), the United Nations General Assembly (UNGA) declared 2023 as the "International Year of Millets". India is a leading producer of millet or nutri-cereals and contributes about 40% to global millet production. Overall, millets provide an excellent opportunity to reshape the nutritional and economic security of India's farmers.

10.4 International Year of Millets (IYM)

The United Nations General Assembly at its 75th session in March 2021 declared 2023 the International Year of Millets (IYM). The IYM 2023 is an opportunity to raise awareness of, and direct policy attention to the nutritional and health benefits of millets and their suitability for cultivation under adverse and changing climatic conditions. It is an occasion to promote the sustainable production of millets while also highlighting their potential to create sustainable market opportunities for producers and consumers. IYM 2023 aims to contribute to the UN 2030 Agenda for Sustainable Development, particularly SDG 2 (Zero Hunger), SDG 3 (Good health and well-being), SDG 8 (Decent work and economic growth), SDG 12 (Responsible consumption and production), SDG 13 (Climate action) and SDG 15 (Life on land).

Briefly given here

1. The sustainable cultivation of millets can support climate-resilient agriculture. SDG 13 (Climate Action) and SDG 15 (Life on Land)
2. The sustainable production of millets can fight hunger and contribute to food security and nutrition: SDG 2 (End Hunger)
3. Millets can be an important part of a healthy diet: SDG 3 (Good Health and Well-Being)
4. Greater consumption of millets can offer opportunities to smallholder farmers to improve their livelihoods: SDG 8 (Decent Work and Economic Growth)
5. Proper handling of millets is key to maintaining their high quality and nutritional benefits: SDG 2 (End Hunger) and SDG 3 (Good Health and Well-Being)
6. Greater trade in millets can improve the diversity of the global food system: SDG 8 (Decent Work and Economic Growth) and SDG 12 (Sustainable Consumption and Production)

10.5 India Millets Production

India is the largest producer of millet in the world. There are two varieties of millet, pearl millet (bajra) and sorghum (sorghum) which together account for about 19% of world production in 2020, with pearl millet producing 40.51% and sorghum 8.09%. The major millet-producing states in India in 2020 are Rajasthan, Karnataka, Maharashtra, Uttar Pradesh, Haryana, Gujarat, Madhya Pradesh, Tamil Nadu, Andhra Pradesh and Uttarakhand. These ten states account for about 98% of the millet production in India during the period 2020–21. Rajasthan's contribution to India's total millet production is 28.61%. Pearl millet (Bajra), jowar (Jowar) and finger millet (Ragi) have the largest share in the total production of millets. The area under cultivation of millets has ranged from 12.29 to 15.48 million hectares from 2013–14 to 2021–22 (Figs. 10.2 and 10.3).

10.6 To Promote Millet, the Following Steps Were Taken by the Government

1. Millet Awareness Quizzes/Competition and Conferences are being conducted by Food Corporation of India and Central Warehousing Corporation of Department of Food and Public Distribution.
2. The Year 2023 is being celebrated as the International Year of Millets-2023 for promoting of millets.

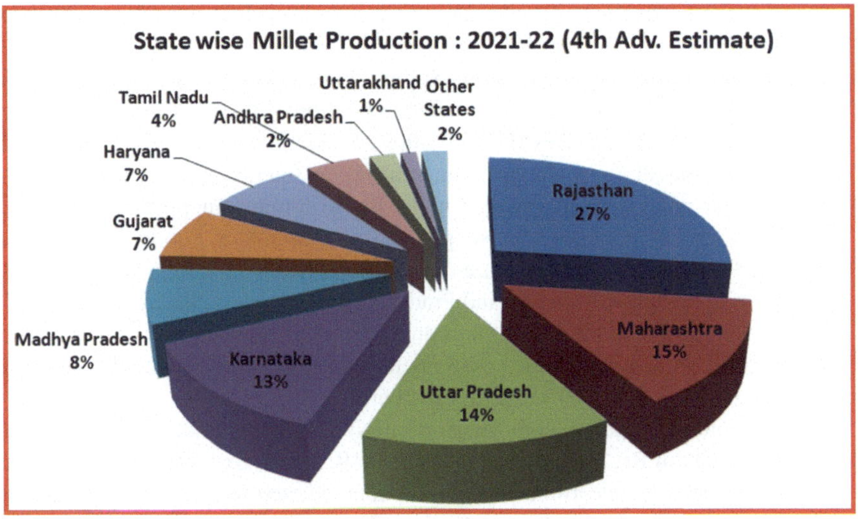

Fig. 10.2 State wise millet production (2021–22). Source: https://apeda.gov.in/

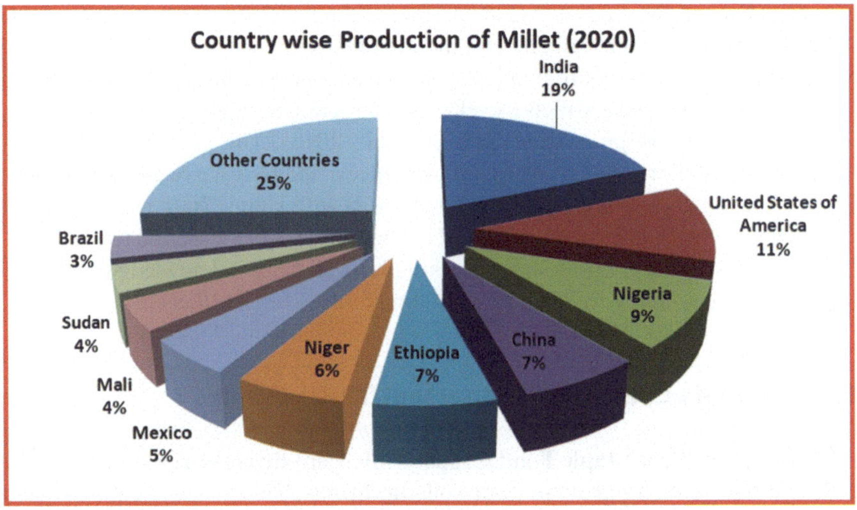

Fig. 10.3 Country wise production of millets (2020). Source: https://apeda.gov.in/

3. Millets have been termed as 'Nutri-cereals' owing to their health benefits.
4. All the offices/Central Public Sector Enterprises of the Department of Food and Public Distribution have been directed to introduce and promote millets in their canteens.
5. National Food Security Mission (NFSM): Launched in October 2007 to ensure holistic development of agriculture and allied sectors The NFSM includes the cultivation of millets as one of its components (Kumar et al., 2018). It aims to

increase the production of millet and other crops to enhance food security and improve the income of farmers.

6. Government of India fixes Minimum Support Prices (MSPs) for twenty-two (22) mandated crops based on the recommendations of the Commission for Agricultural Costs & Prices (CACP) after considering the views of concerned State Governments and Central Ministries/Departments. The millets such as Ragi, Jowar and Bajra are covered under the Minimum Support Price (MSP). The government allocates coarse grains (millets) under Public Distribution System (PDS) and Other Welfare Schemes (OWSs) including Wheat Based Nutrition Programme (WBNP) and Pradhan Mantri Poshan Shakti Nirman (PM POSHAN) Schemes based on the requirement projected by the State Govts., Ministry of Women and Child Development and Department of School Education & Literacy subject to the availability of coarse grains (millets).
7. Nutri-Cereals Campaign: Sub-Mission on Nutri Cereals under NFSM started with an outlay of Rs. 300.00 crores for 2018–19. It aims to develop a strategy for addressing issues concerning production, demand, and research with a market-oriented approach. Increasing production of Nutri Cereals through area expansion and productivity enhancement in a sustainable manner in the identified districts of the country. Strengthening seed supply system of Nutri-Cereals. It was enhancing postharvest value addition at the farm gate for better price realization to farmers through efficient market linkages.
8. Millets Development Program: The Indian Council of Agricultural Research (ICAR) has initiated a Millets Development Program to support millets' research, development, and promotion (Saxena et al., 2018). The program focuses on improving millet varieties, agronomic practices, and post-harvest technologies.
9. NITI Aayog signed a Statement of Intent (SoI) with United Nations World Food Program (WFP) on December 20, 2021. The partnership focuses on mainstreaming millets and supporting India in taking lead globally in knowledge exchange using the opportunity of 2023 as an International Year of Millets.

10.7 Millets Considered Important 'Nutri-Cereals'?

- **Climate Resilient Staple Food Crops:** Millets are **drought-resistant, require less water and can grow in poor soil conditions.** This makes them a suitable food crop for areas with unpredictable weather patterns and water scarcity.
- **Rich in nutrients:** Millets are a good source of fiber, protein, vitamins, and minerals.
- **Gluten-free:** Millets are naturally gluten-free, making them suitable for people with celiac disease or gluten intolerance.
- **Adaptable:** Millets can be grown in a variety of soils and climates, making them a versatile crop option for farmers.
- **Sustainable:** Millets are **often grown using traditional farming methods**, which are more sustainable and environmentally friendly than modern, industrial farming practices (Sharma et al., 2021).

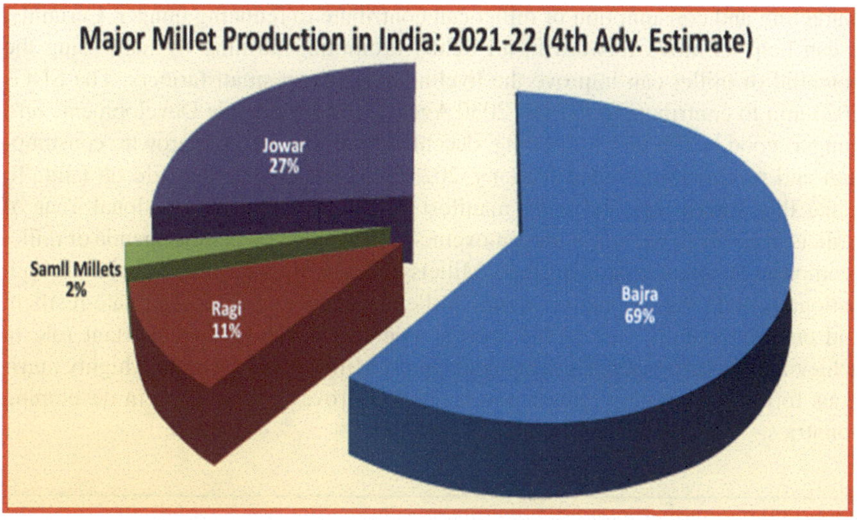

Fig. 10.4 Major millet production in India (2021–22). Source: https://apeda.gov.in/milletportal/Production.html

According to the available data, India's millet production has shown a positive trend in recent years. India produces around 12 million MT of millets annually, according to Ministry of Agriculture and Farmers Welfare data. Pearl millet (bajra) is the most extensively cultivated millet variety in India, followed by Sorghum (Jowar), ragi and small millet (Fig. 10.4).

10.8 Why Millet?

Millets like jawar, pearl millet (bajra), and finger millet (ragi) are known as nutrient-rich superfoods. Millets are capable of filling people's stomachs without causing any harm to the environment. Millet is popular not only for its health benefits but also for its environmental impact. It is low in calories and rich in fiber, which helps in reducing weight. Millets are rich in antioxidants that protect your cells, and do not spike blood sugar, which is good for people with diabetes. Millet is an excellent source of dietary fiber, which aids digestion. The fiber present in millet acts as a prebiotic, which nourishes the beneficial gut bacteria. It is naturally low in saturated fat and is known to increase the levels of bad LDL cholesterol in the blood.

10.9 Concluding Remark

Millets contain 60–70% carbohydrates, 7–11% proteins, 1.5–5% fat, and 2–7% crude fibre and are also rich in vitamins and minerals. The declaration of the International Year of Millet recognizes that sustainable production, processing, marketing and consumption of millet can contribute to reducing hunger. Certainly, it can help us move forward in all fields. Promoting farming by harnessing the potential of millet can improve the livelihood of many small farmers. The SDGs 2023 aim to contribute to the UN 2030 Agenda for Sustainable Development, zero hunger, good health and well-being, decent work and economic growth, consumption and production (United Nations, 2023), climate change and life on land. To make this sustainable and truly manifest the spirit of the International Year of Millets, it needs to become a mass movement. IYM 2023 places the agenda of millet promotion on an international stage. Millets will now re-enter the food platters both nationally and internationally. Overall, millet is a highly nutritious, climate-resilient and environmentally sustainable food source that can play an important role in achieving the Sustainable Development Goals. Millet is a low-cost and highly nutritious food that can help reduce poverty and improve food security in developing countries.

References

Banerjee, P., & Maitra, S. (2020). The role of small millets as functional food to combat malnutrition in developing countries. *Indian Journal of Natural Sciences, 10*(60), 20412.

Drishti IAS. (2023). *International year of millets*. www.drishtiias.com/daily-updates/daily-news-analysis/international-year-of-millets-2. (Accessed on 6th June 2023).

https://apeda.gov.in/apedawebsite/SubHead_Products/Indian_Millets.htm

https://apeda.gov.in/milletportal/Production.html

https://pib.gov.in/newsite/erelcontent.aspx?relid=106818

https://pib.gov.in/PressReleaseIframePage.aspx?PRID=1796559

https://pib.gov.in/PressReleasePage.aspx?PRID=1783716

https://www.nutricereals.dac.gov.in/IYoM2023/Data/Millets%20International%20Recipes.pdf

Kumar, A., Tomer, V., Kaur, A., Kumar, V., & Gupta, K. (2018). Millets: A solution to agrarian and nutritional challenges. *Agriculture & Food Security, 7*.

Mounika, M., & Uma Devi, K. (2019). Dietary fibre content in healthy millet based recipes. *International Journal of Current Microbiology and Applied Sciences, 8*, 1829–1834.

Saxena, R., Vanga, S. K., Wang, J., Orsat, V., & Raghavan, V. (2018). Millets for food security in the context of climate change: A review. *Sustainability, 10*.

Sharma, R., Sharma, S., Dar, B. N., & Singh, B. (2021). Millets as potential Nutri-cereals: A review of nutrient composition, phytochemical profile and techno functionality. *International Journal of Food Science and Technology, 56*, 3703–3718.

United Nations. (2023). Transforming our World: The 2030 agenda for sustainable development.

YOJANA January 2023, Volume-67 No. 01.

Nutritional Security Through Millet Value Chains: Indian Success Story

11

Priyanshu Pal, Manas Vijay, and Shubhangi Nigam

Abstract

Millets are often referred to as "nutri-cereals" due to their high nutritional value, resilience to climate change, and suitability for marginal environments. In India, the revival of millet cultivation and its value chain presents a promising pathway to enhance nutritional security, particularly for vulnerable populations. This research explores the role of millet value chains in addressing malnutrition, promoting sustainable agriculture, and enhancing farmers' livelihoods. Through a comprehensive analysis of policy frameworks, production systems, market dynamics, and consumption patterns, this study provides insights into how millets can contribute to food security and sustainable development in India.

Keywords

Millets · Nutritional security · Value chains · Sustainable agriculture · India · Food security · Malnutrition

11.1 Introduction

In general, food security of a nation refers to a situation when all people, at all times, have physical, social, and economic access to sufficient, safe, and nutritious food that meets their dietary needs and food preferences for an active and healthy life (Food and Agriculture Organization, 1996). A deficiency of any component of this condition is an indication of food insecurity. Greater deficit increases the severity of food insecurity. The food and agriculture organization (Given the increase of population, climate change, and nutrition requirement, India is faced with a need for

P. Pal (✉) · M. Vijay · S. Nigam
Department of Food Technology, Institute of Engineering and Technology, Bundelkhand University, Jhansi, UP, India

C. B. Singh et al. (eds.), *Global Millets Production for a Sustainable Future*,
https://doi.org/10.1007/978-3-031-88157-2_11

Food and Agriculture Organization, 2013) defined food insecurity as a situation of insufficient access to safe and healthy food. The dimension of food security not only includes stable access to physical availability of food, but also sustained access to appropriate nutrition. The Food and Agriculture Organization (FAO) begins segregating food insecurity into moderate (lack of consistent access to food resulting diminished dietary quality, eating patterns, health, and wellbeing) and severe (running out of food, experiencing hunger, and having health and wellbeing at severe risk) security (Food and Agriculture Organization, 2019). However, nutrition security alone is only implied in the definition of food security without any specific focus. The UN Standing Committee on Nutrition integrates nutrition security explicitly in the situation of food security and coined the term food and nutrition security, which defines the condition when all people at all times have physical, social and economic access to food, which is consumed in sufficient quantity and quality to meet their dietary needs and food preferences. This condition is supported by an environment of adequate sanitation, health services and care, allowing for a healthy and active life 'In recent years, nutrition security, has gained importance over food security'. Ingram suggests using the term nutrition security, as he argues that nutrition security encompasses food security (Ingram, 2020). Later, the worldwide effort of Green Revolution (GR) to produce cereal foods, especially rice and wheat, to combat the calorie requirement of the increasing population played a significant role in combating food security. Despite considerable improvement in increased food production and improved distribution, ensuring enough food with appropriate nutrition remains elusive for all people at all times. Hunger and malnutrition continued in the world, including India. Although considerable progress has been made in India in improving food and nutrition security, there is still a long way to go new direction and policy to achieve food and nutrition security. On one hand, as a policy perspective, India introduced The National Food Security Act (also Right to Food Act) in 2013. The Act aims to provide subsidized food grains to approximately two-thirds of India's 1.3 billion people. On the other hand, for diversification of food production and grain-mix in the PDS, India has given importance to millets. Millets, a group of hardy and drought-resistant cereal crops, have been incorporated in the cropping system to improve food and nutritional security. In this study, the potential of millets to combat food and nutrition insecurity was explored in India with special focus on what determines its contribution and what policy actions are necessary.

11.2 Literature Review

11.2.1 Nutritional Profile of Millets

Millets are rich in protein, fiber, vitamins, and minerals. Table 11.1 summarizes the nutritional content of various millet types compared to rice and wheat.

Table 11.1 Nutrional profile of millets

Nutrient (per 100 g)	Foxtail Millet	Pearl Millet	Rice	Wheat
Protein (g)	12.3	11.6	2.7	12.6
Fiber (g)	8.0	7.0	0.4	2.7
Iron (mg)	2.8	3.0	0.2	0.6
Calcium (mg)	17	42	10	15

11.2.2 Historical Context of Millet Cultivation in India

Millet cultivation has a long history in India, with diverse cultural significance. However, policies favoring major cereals have marginalized millets.

11.2.3 Current Trends in Millet Production

Recent trends show a gradual revival in millet cultivation, driven by health awareness and government initiatives.

11.3 Food and Nutritional Security in India

11.3.1 Millets as Contributors of Nutritional Security in India

Millets are a group of diverse small-grain cereal crops grown in marginal soil and under stressed conditions. They comprise about a dozen crop species originated in Asia and Africa, primarily in the third world countries (Gupta et al., 2017). India is the largest producer of millets producing nearly 40% of the world's millets despite much negative pressure from competing crops in terms of policies and production supports. Among several types of millets (pearl millet, finger millet, proso millet, foxtail millet, guinea millet, little millet, barnyard millet, etc.), pearl millet is the most common in India accounting for nearly 75% of the total area devoted to millets (Bhagavatula et al., 2013). The millets ability to adapt to adverse climatic conditions, requirement of minimal inputs, and superior nutritional qualities are among the specific characteristics of millets rarely found in other common cereals. Millets possess specific molecular, biochemical, and morpho-physiological characteristics that allow them to withstand adverse environmental conditions–drought and poor soil conditions (Bandyopadhyay et al., 2017). Their shorter lifecycle, short stature, and small leaf area may also offer an added advantage. Considering their adaptability to high light, high temperature, and dry weather as C4 photosynthetic capacity, millets are more efficient than common cereal crops. These mechanisms allow millets to have enhanced photosynthetic rates and lower photorespiration rates, as well as water and nitrogen use efficiency under warm conditions (Bandyopadhyay et al., 2017). The millet producing areas of India as shown in Fig. 11.1 are on the western part, which are susceptible to drought. Millets are important sources of nutrients and can play a significant role in improving nutritional security and preventing

Fig. 11.1 Food insecurity and food security

diseases caused by imbalanced nutrition. They are gluten-free and contain as much protein as wheat does. In terms of macronutrients, millets are either similar or superior to major cereals (Kumar et al., 2018). They also contain several micronutrients, vitamins, insoluble dietary fiber, and phenolic compounds, which are essential for health benefits. They are thought to have several health benefits including the ability to address diabetes, aging, cancer, celiac disease, and cardio-vascular disease (Bhat et al., 2018). Millet-based health food items are common and exhibit longer storage life. Durairaj et al. (2019) observed a significant increase in height, weight, and hemoglobin level of the school children who regularly consumed millet-based health food (Durairaj et al., 2019). Even though millets are produced in adverse climatic conditions and can provide nutrition that otherwise are hard to find, their acreage continues to decline, and production and yield have stagnated. Between 1966 and 2006, India lost 44% of millet cultivation areas to other crops due to lack of policy support.

Given that the nutritional security is a global phenomenon, India has joined the international partners and governments to implement policies and regulations that would offer support to the NUSs including millets. The Seeds Act 1966, the Patent Act 1970, the Protection of Plant Variety and Farmers, Rights Act 2002, and the Biological Diversity Act 2002 are some of the regulations that offer support to the continuing development of the supply side of millets with appropriate implementation. Similarly, appropriate incorporation of millets in the PDS, along with applicable MPS, can improve the demand side of millets (Table 11.2).

Table 11.2 Food and nutrition security in India (figures are on averages for three consecutive years, except for percent women anemic, which is for the middle year) Average of 3 years

Average of 3 years	Dietary energy supply (%)	Value of rood production ($/ capita)	Dietary energy from cereals, roots. Tubers (kcal/cap/d)	Protein supply (g/cap/d)	Animal protein (g/cap/d)	Percent undernourished	People undernourished (mill)	women anemic (15–49 years) %
1999–01	103	150	63	56.3	9.0	18.2	191.2	53.3
2000–02	101	144	63	55.3	9.0	18.6	199.6	53.3
2001–03	100	145	62	54.7	9.0	20.2	220.7	53.3
2002–04	99	142	62	54.0	9.0	21.7	240.7	53.3
2003–05	100	146	62	53.7	9.3	22.2	251.1	53.3
2004–06	101	148	62	54.0	9.7	21.7	249.4	53.2
2005–07	103	156	61	55.0	10.3	19.8	230.3	53.0
2006–08	104	162	60	56.7	10.7	17.6	208.7	52.7
2007–09	104	164	60	57.3	11.0	16.7	200.6	52.4
2008–10	104	165	59	58.0	11.0	16.4	199.6	52.0
2009–11	104	169	59	58.7	11.3	16.3	200.9	51.7
2010–12	105	176	58	59.3	11.7	16.3	203.6	51.5
2011–13	105	181	58	59.7	11.7	16.3	206.4	51.3
2012–14	106	184		60.0	11.7	15.9	203.8	51.2
2013–15	106	185		60.7	12.0	15.3	198.3	51.1
2014–16	107	186		61.7	13.0	14.7	193.1	51.2
2015–17	108			63.0	13.7	14.4	190.9	51.4
2016–18	109					14.2	190.1	
2017–19						14.0	189.2	

Source: FAO http://www.fao.org/faostat/en/#data/FS

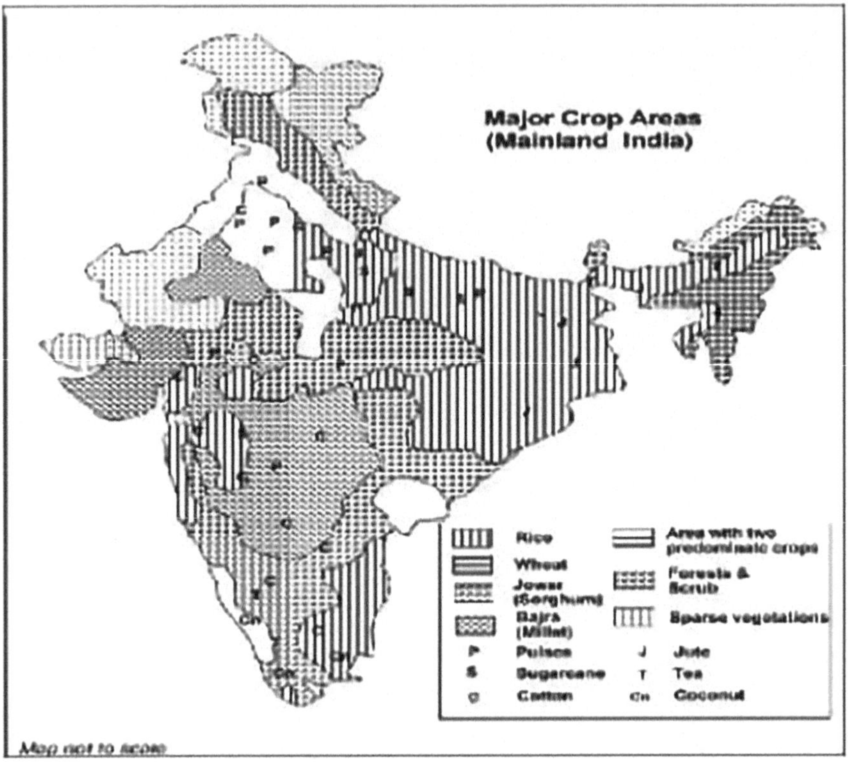

Fig. 11.2 Major cropping regions of India (Gurditta & Singh, 2016)

A mixed-method approach combining quantitative data analysis and qualitative interviews with farmers, traders, and consumers was employed.

11.3.2 Data Collection

Data were collected through surveys, focus group discussions, and secondary sources from government reports and academic publications (Figs. 11.2 and 11.3).

11.3.3 Data Analysis

Statistical tools were used to analyze production data, while thematic analysis was applied to qualitative data.

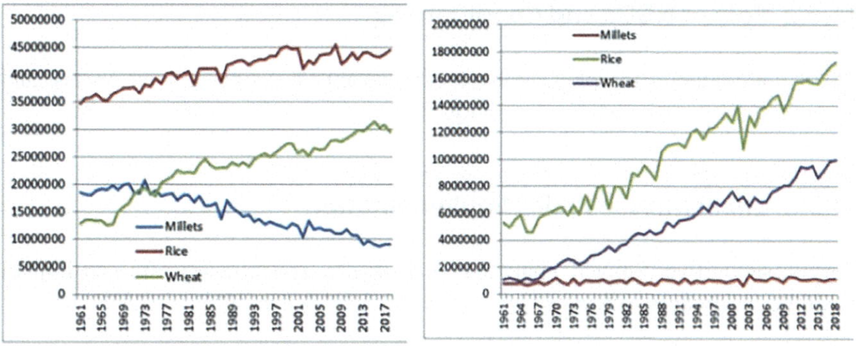

Fig. 11.3 Average and production of millet, rice, and wheat (Left: Acreage in hectares; Right: Production in tons)

11.4 Millet Value Chains in India

11.4.1 Overview of the Value Chain

The millet value chain includes production, processing, marketing, and consumption. Figure 11.1 illustrates the stages of the millet value chain.

11.4.2 Stakeholders Involved

Key stakeholders include farmers, processors, retailers, and consumers, each playing a critical role in the value chain.

11.4.3 Market Dynamics

Analysis of market prices, demand trends, and supply chain logistics reveals opportunities for improving millet marketing.

11.5 Policy Framework and Interventions

11.5.1 Government Policies Supporting Millets

Several policies promote millet cultivation, including the National Food Security Mission and the Pradhan Mantri Krishi Sinchai Yojana.

11.5.2 Role of NGOs and Cooperatives

NGOs and cooperatives play vital roles in capacity building, market access, and promoting consumer awareness about millets.

11.5.3 Challenges in Policy Implementation

Despite supportive policies, challenges like inadequate infrastructure and limited consumer awareness persist.

11.6 Case Studies of Successful Millet Initiatives

11.6.1 State-Level Success Stories

This chapter highlights successful millet initiatives from various Indian states, focusing on innovative practices and community engagement.

11.6.2 Impact Assessment

Assessing the socio-economic impact of these initiatives shows improved livelihoods for farmers and increased nutritional intake for consumers.

11.6.3 Lessons Learned

Key lessons from these case studies emphasize the importance of collaboration among stakeholders and sustainable practices.

11.7 Conclusion and Recommendations

It was realized that millets have substantial potential to contribute toward food and nutritional security in India. As a result, millets should be brought out of NUS and considered as another staple food along with rice and wheat. Accordingly, research and development efforts and policy formulations are required; some steps have been taken throughout the world, especially in India. However, these efforts are far from adequate. On the supply side, appropriate implementation of the relevant regulations is necessary to ensure continued research and development for improved varieties, availability of quality seeds, adequate support for cultivation, satisfactory technology for processing, and marketing millets. On the demand side, millets should be included in the PDS along with rice and wheat, so that they appropriate

MPS. At the same time, focused marketing strategies and product development of new and better millet-based products are needed.

11.7.1 Summary of Findings

Millets have the potential to significantly enhance nutritional security in India through a well-structured value chain.

11.7.2 Recommendations for Policy and Practice

Strengthen infrastructure for millet processing and marketing.

Increase consumer awareness through education and promotional campaigns.

Support research and development for improved millet varieties and sustainable farming practices.

11.7.3 Future Research Directions

Further research should focus on long-term impacts of millet value chains on nutritional outcomes and environmental sustainability.

References

Bandyopadhyay, T., Muthamilarasan, M., & Prasad, M. (2017). Millets for next generation climate-smart agriculture. *Frontiers in Plant Science, 8*, 1266.

Bhagavatula, S., Rao, P. P., Basavaraj, G., & Nagaraj, N. (2013). *Sorghum and millet economies in Asia–Facts, trends and outlook*. International Crops Research Institute for the Semi-Arid Tropics.

Bhat, S., Nandini, C., & Tippeswamy, V. (2018). Significance of small millets in nutrition and health—A review. *Asian Journal of Dairy & Food Research, 37*(1).

Chander, S., Vermani, S., & Kumar, A. (2017). Role of public distribution system in providing food security in India. *Indian Journal of Health and Wellbeing, 8*(4), 322.

Durairaj, M., Gurumurthy, G., Nachimuthu, V., Muniappan, K., & Balasubramanian, S. (2019). Dehulled small millets: The promising nutricereals for improving the nutrition of children. *Maternal & Child Nutrition, 15*, e12791.

Food and Agriculture Organization. (1996). Rome declaration on world food security and world food summit plan of action. In *World Food Summit, 13–17 November 1996*. Food and Agriculture Organization of the United Nations, Rome.

Food and Agriculture Organization. (2013). *The state of food insecurity in the world, 2013: The multiple dimensions of food security*. FAO.

Food and Agriculture Organization. (2019). *The state of food security and nutrition in the world: Safeguarding against economic slowdowns and downturns*. FAO.

George, N. A., & McKay, F. H. (2019). The public distribution system and food security in India. *International Journal of Environmental Research and Public Health, 16*(17), 3221.

Gupta, S. M., et al. (2017). Finger millet: A "Certain" crop for an "Uncertain" future and a solution to food insecurity and hidden hunger under stressful environments. *Frontiers in Plant Science, 8*, 643.

Gurditta, H., & Singh, G. (2016). Climate change, food and nutritional security: Issues and concerns in India. *Journal of Climate Change, 2*(1), 79–89.

Ingram, J. (2020). Nutrition security is more than food security. *Nature Food, 1*(1), 2–2.

Khairwal, I., & Improvement, A. I. C. P. M. (2005). Pearl millet (Pennisetum glaucum) improvement in India-retrospect and prospects. *Indian Journal of Agricultural Sciences, 75*(4), 183–191.

Kumar, A., Tomer, V., Kaur, A., Kumar, V., & Gupta, K. (2018). Millets: A solution to agrarian and nutritional challenges. *Agriculture & Food Security, 7*(1), 1–15.

Pillay, D., & Kumar, T. M. (2018). Food security in India: Evolution, efforts and problems. *Strategic Analysis, 42*(6), 595–611.

Rai, R. K., Kumar, S., Sekher, M., Pritchard, B., & Rammohan, A. (2015). A life-cycle approach to food and nutrition security in India. *Public Health Nutrition, 18*(5), 944–949.

Ritchie, H., Reay, D., & Higgins, P. (2018). Sustainable food security in India—Domestic production and macronutrient availability. *PLoS One, 13*(3), e0193766.

Sarkar, A. (2015). Modern agriculture and food and nutrition insecurity: Paradox in India. *Public Health, 129*(9), 1291–1293.

Sharma, S., & Sharma, P. (2017). Agricultural production, marketing and food security in India: A peep into progress. *Productivity, 58*(2).

Varadharajan, K. S., Thomas, T., & Kurpad, A. V. (2013). Poverty and the state of nutrition in India. *Asia Pacific Journal of Clinical Nutrition, 22*(3), 326–339.

Food Security and Climate Change in India

12

Debesh Bhowmik

Abstract

In this paper author tried to verify the relationship between climate change and food security in India during 1990–2023 applying cointegration and vector error correction model where temperature change, emission from agricultural land use, average rainfall, emissions from CO_2 and emission from methane are considered as determinants of climate change and food supply in kilo calorie per capita per day is considered as indicator of food security. The paper observed that average rainfall, emissions from CO_2 and emission from methane have significant long run causality with food supply in kilo calorie per capita per day which has also short causality with temperature change and average rainfall. The cointegrating relationships converge towards equilibrium significantly.

Keywords

Food security · Emissions · Food supply · Cointegration · Vector error correction · Short run causality · Long run causality

JEL Classification Codes

C22 · F53 · F55 · I32 · O13 · Q10 · Q18 · Q51 · Q54 · Q58

D. Bhowmik (✉)
Faculty-Business and Accountancy, Lincoln University College,
Petaling Jaya, Selangor, Malaysia

C. B. Singh et al. (eds.), *Global Millets Production for a Sustainable Future*,
https://doi.org/10.1007/978-3-031-88157-2_12

12.1 Introduction

Food security and livelihood are both affected by climate change and the affected people fell into hunger since food production was hindered due to climate impacts and it is fact that at present, 25.9% global population have experienced hunger having without access to nutritious and sufficient food in 2019 (FAO, 2020). The functioning of all the components of food systems were affected by climate change (IPCC, 2019) where food system contributes 21–37% emissions of GHG from agricultural land use, storage, transport, packaging, processing, retail, and consumption (Mbow et al., 2019).

An increase of 1–2 °C in temperature is going to have a significant negative impact on the yield of major food crops in low-altitude countries like India (Bhat & Pandit, 2020). In India, undernourished people stood at 197.2 million in 2000–2002 which catapulted to 233.9 million in 2020–2022. Value of food imports of India in total merchandise exports (percent) (3-year average) was 5% in 2000–2002 which increased to 6% in 2019–2021. India's food supply in rice in kg/per man per year was only 102.91 kg in 2010 which marginally rose to 104.78 kg in 2021 only. India's food price inflation rate was 2.99% in 2001 which reached its peak level in 2020 of 11.71% which now stood at 6.189% in 2023 January which is too high (FAO). It was estimated that about 40.6% of India's population are moderate or severe food insecure in 2019–21. The severely food insecure population has risen from 20.3% in 2018–20 to 22.3% in 2019–21 (Bansal, 2022).

In brief, climate change calamities disrupted food system, availability of food and supply chain of food distribution and agricultural production which finally hinders the food security for which poverty, hunger and inequality aggravate systematically and ultimately affects sustainable development.

In this paper, author tried to examine the impact of climate change on food security in India during 1990–2023 through econometric model of Johansen cointegration and vector error correction analysis.

12.2 Literature Review

There are many researches on climate and food security in global or regional or country level studies. In this paper, author used some of the important literatures which are related to this study.

Khan and Hassan (2017) showed by statistical evidences in India that the temperature change, average rainfall and emission of CO_2 have affected significantly on agriculture and thus on food security.

Kaur and Kaur (2018) verified the impact of climate change on food security by using linear regression during 1990–2014 in India and found that a rise in average minimum temperature led to reduce agricultural food index and employment in agriculture. Besides, a rise of rainfall led to reduce agricultural production index and depth of food deficit significantly.

Jeder et al. (2020) examined in Tunisia during 1991–2017 by applying cointegration and VECM and found that temperature change has adverse effect on food security while food import, land under cereals and inflation have significant negative relation with climate change in the short run. Climate change and population are neither significant nor important to explain Tunisia's food security.

Singh, Anand and Singh (2020) studied the impacts of climate change on crop yields in India in Himalayan and Gangetic plains during 1966–2011 and observed that in western Himalayan region, temperature change affects negatively on rice and wheat and rainfall negatively impacts on wheat and oilseed while in eastern region, only rainfall adversely induces on oilseed but temperature rise affects negatively on rice, wheat and oil seed. In Gangetic plains, temperature negatively impacts on rice, wheat and oilseed but rainfall only affects oilseed significantly.

Kumar and Sharma (2022) studied state level analysis in India during 1980–2009 and concluded that climate change affects foodgrain and non-foodgrain productivity negatively which impacts on food security.

Dasgupta and Robinson (2022) examined in 83 countries between food insecurity and climate data applying panel data regression analysis taking data from FAO and ERAS-Land and found that 1 °C temperature increase led to 1.4% increase in severe global food insecurity in 2014 compared to 1.64% increase in 2019.

Affoah et al. (2022) studied 25 sub-Saharan African countries from 1985 to 2018 by applying ARDL approach to ensure pool mean group using modified ordinary least squares and dynamic ordinary least squares regression and found that rainfall and CO_2 emission have a significant positive effect on food availability in the long run while temperature's effect is significantly negative. Temperature has an inverse impact on food accessibility and rainfall also has a positive influence on food utilization.

Bhuyan et al. (2023) tested the influence of climate change on food security in India during 1994–2019 applying Autoregressive Distributed Lag method and found that a 1% increase in area under cultivation led to a 2.4% increase in food security in the short run and a 3.2% increase in the long run or, 1% increase in population growth will decline food security by 3.8% in the short run and 7.8% in the long run. Besides, 1% temperature increase leads to a decrease of food security by 1.2% in the short run and 1.7% in the long run.

12.3 Objectives

The paper endeavours to verify the impact of climate change indicators namely, temperature change, emission from agricultural land use, average rainfall, emissions from CO_2 and emission from methane on the food security indicator viz. food supply in kilo calorie per capita per day in India from 1990 to 2023 in India utilizing the cointegration and vector error correction model.

12.4 Methodology and Data

Johansen cointegration test (1988) can be expressed as follows.

If the m × 1 vector time series Y_t contains more than 2 components, each being I (1), then there may exist k(<m) linearly independent 1 × m vectors $\alpha_1, \alpha_2, \ldots, \alpha_k$ such that $\alpha' y_t \sim I(0)$ k × 1 vector process where $\alpha = (\alpha_1, \alpha_2, \ldots, \alpha_k)$ is a k × m cointegrating matrix.

Let VAR(p) [Vector Auto Regressive] model is given below,

$$Y_t = \delta D_t + \phi_1 Y_{t-1} + \ldots\ldots\ldots\ldots + \phi_p Y_{t-p} + \varepsilon_t$$

Where Y_t is a time series m × 1 vector of I (1) variables. The VAR(p) model is stable if

$$\text{Determinant}\left(\text{In-}\phi_1 z - \ldots\ldots\ldots\ldots - \phi_p z^p\right) = 0$$

If there are roots on the unit circle then some or all the variables in Y_t are I (1) and they may be cointegrated. If cointegration exists the VAR model is transferred to VECM (Vector Error Correction Model) which is given below.

$$\Delta y_t = \Gamma_0 D_t + \Pi Y_{t-1} + \sum_{j=1}^{p-1} \Gamma j \Delta Y_{t-j} + \varepsilon_t$$

Where D_t = vector deterministic variables (constants, trends, and/or seasonal dummy variables).

$$j = -I + \phi_1 + \ldots\ldots\ldots\ldots \phi_j, \text{for all values } j = 1,2,\ldots\ldots\ldots p-1 \text{ are } m \times m \text{ matrix.}$$

$\Pi = \gamma A$ is the long run impact matrix, A and γ are mxk matrices,

$$\varepsilon_t \text{ are } Nm(0,\Sigma) \text{ errors.}$$

Determinant $(1 - \sum_{j=1}^{p-1} \Gamma j B^j)$ has all its roots outside the unit circle.

Assume VECM errors are independent Nm (0, Σ) distribution, then given CI restrictions on the trends/drift/no drift parameters, the likelihood $L_{max}(k)$ is a function of the CI rank k.

The Trace test is based on log-likelihood ratio (LR).

$$LR = 2\ln\left[L_{max}(\text{unrestricted}) / L_{max}(\text{restricted})\right] \text{for } k = m-1,\ldots\ldots1,0.$$

The Test H_0:CI rank ≤k Vs H_1:CI rank>k. If the LR is greater than the critical value for a certain rank, then H_0 is rejected.

The Trace Test: $LR_{trace}(k) = -2\ln\Lambda = -T\sum_{i=k+1}^{m} \ln(1-\lambda i)$ where λ_i denotes the descending ordered Eigen values $\lambda_1 > \lambda_2 > \ldots\ldots > \lambda_m > 0$ of the determinant $(\lambda S_{11} - S_{10} S00^{-1} S_{01}) = 0$.

If $LR_{trace}(k)$ > CV (for rank k), then H_0(CI rank k) is rejected.

Alternatively,

$$LR_{max}(k) = -2\ln\Lambda = -T\ln(1-\lambda_{k+1}) \text{ is called the maximal Eigen value statistic.}$$
$$\text{Test } H_0: CI\,rank = k \text{ Vs } H_1: CI\,rank = k+1.$$

The impulse response function can be expressed as MA (Moving Average) process which is derived from VAR(p) model as,

$$Y_t = c + \varepsilon_t + \psi_1 \varepsilon_{t-1} + \psi_2 \varepsilon_{t-2} + \ldots\ldots \psi(B)\varepsilon_t$$

$$\Delta y_{t+1} / \delta\, \varepsilon_t = \psi_1$$

The row i, column j element of ψ_l identifies the consequences of one unit increase in the jth variable's innovation at date $t(\varepsilon_{tj})$ for the value of the ith variable at time t + 1, holding all other innovations at all dates constant. A plot of the row i, column j element of as a function of lag l is called the non-orthogonalized impulse response function.

The Wald test (1943) has been explained in brief which is relevant in this analysis.

Y_t in a VAR(p) process is divided into sub-processes z_t and x_t i.e., $\ddot{Y}_t = (\dot{x}_t, \dot{z}_t)$. Causality is defined by zero constraints on VAR coefficients and when in VAR(p) system, it is tested zero constraints for the coefficients to derive asymptotic test considering ca = c < = >A_{jki} = 0 in a manner of:

$$H_0 : ca = c \text{ against } H_1 : ca \neq c$$

where ca is an $(N\times(k^2p + k)$ matrix of rank N and c is an $(N\times1)$ vector. Assuming that $\sqrt{T}(c\ddot{a} - a)\cap N[(0, \Gamma^{-1}\Sigma_u)']$ is a least square/maximum likelihood estimation and then we get $\sqrt{T}(c\ddot{a} - ca) \cap N[0, c(\Gamma^{-1}\Sigma_u)c']$. Hence the Wald statistic is given by

$$T(c\ddot{a} - c)'\left[c\left(\Gamma^{-1}\Sigma_u\right)c'\right]^{-1}(c\ddot{a} - c) \cap \chi 2(N).$$

Then we replace Γ and I_u by their unusual estimator $\Gamma' = zz'/T$ and $\Sigma'_u = T/(T-k_p-1)$ $x\Sigma'_u$, then the resulting statistic becomes $\lambda_w = (c\ddot{a} - a)'[c((zz')^{-1}\,\Sigma'_u)c']^{-1}(c\ddot{a} - c)$.

It is still asymptotic $\chi 2$ distribution with N degree of freedom and it has the condition of $[c((zz')^{-1}\,\Sigma'_u)c']^{-1}/T$ which is a consistent estimator of $[c(\Gamma^{-1}\Sigma_u)c']^{-1}$.

Hence, we have the following result: The asymptotic distribution of the Wald statistic supposed.

$\sqrt{T}\,(c\ddot{a}\ \text{-}a)\cap N[(0, \Gamma^{-1}\Sigma_u)]$ holds and in practice, $NF(N,T) \cap \chi 2(N)$ as t tends to infinity where F(N,T) indicates as F random variable with N and T degrees of freedom because F(N,T) distribution has flatter tail than the $\chi 2(N)/N$ distribution and it is reasonable to consider the test statistic $\lambda_F = \lambda_w/N$ in conjunction with critical values from the F -distribution.

The data on temperature change in Celsius degree during 1990–2023, emissions per area of agricultural land in ton/hectares from 1990–2023 of India were collected from FAO, and data on total CO_2 emissions in Kt and methane emissions in kt from 1990 to 2023 of India were collected from the World Bank. And data on food supply in kilo calorie per capita per day of India from 1990 to 2023 were collected from the World Bank data website. Data on average rainfall in milli meter of India during 1990–2023 were collected from the Meteorology department of India.

12.5 Observations and Results

The variables of food supply in kilo calorie per capita per day(y), temperature change in °C(x_1), emissions from agricultural land use in ton/hectare(x_2), average rainfall in milli metre(x_3), CO_2 emission in kt(x_4) and methane emission in kt(x_5) for India from 1990 to 2023 have unit roots but their first difference series contain no unit root which were verified by ADF test statistic which are given below in Table 12.1.

Johansen cointegration test (1988) among the first difference series of food supply in kilo calorie per capita per day(y), temperature change in °C(x_1), emissions from agricultural land use in ton/hectare(x_2), average rainfall in milli metre(x_3), CO_2 emission in kt(x_4) and methane emission in kt(x_5) for India from 1990 to 2023 in all categories have been shown in Table 12.2 which established that there are at least three cointegrating equations among them.

Therefore, the test on quadratic intercept and trend unrestricted cointegration rank test during 1990–2023 (adjusted 1993–2023 for cointegration) of the first difference series among those variables states that Trace statistic has 4 cointegrating equations and Max-Eigen statistic has 3 cointegrating equations which are shown in Table 12.3.

Since, the variables are cointegrated then the estimation of vector error correction model is necessary which is given below in a tabular form (Table 12.4).

Table 12.1 Unit root test

Variable	ADF statistic	Probability	H0 = has unit root
y	1.257901	0.6370	Accept
dy	4.791094	0.0005	Reject
x_1	6.864428	0.0000	Reject
dx_1	10.89201	0.0000	Reject
x_2	1.276695	0.6285	Accept
dx_2	5.630646	0.0001	Reject
x_3	0.668280	0.8407	Accept
dx_3	8.95797	0.0000	Reject
x_4	0.349789	0.9065	Accept
dx_4	4.396120	0.0015	Reject
x_5	1.890754	0.3324	Accept
dx_5	3.796413	0.0070	Reject

Source: Calculated by author, d = first difference

Table 12.2 Summery results of cointegration test

Data trend	none	none	linear	linear	quadratic
Test type	No intercept, no trend	Intercept and no trend	Intercept, no trend	Intercept, trend	Intercept, trend
Trace statistic	3	3	4	3	4
Max Eigen statistic	3	3	3	1	3

Source: Calculated by author, Critical values based on MacKinnon et al. (1999) at 5% level

Table 12.3 Johansen cointegration test

Hypothesized No. of CE(s)	Eigenvalue	Trace Statistic	0.05 Critical Value	Probability**
None*	0.815524	159.4849	107.3466	0.0000
At most 1*	0.699349	107.0876	79.34145	0.0001
At most 2*	0.631478	69.83158	55.24578	0.0016
At most 3*	0.518533	38.88568	35.01090	0.0183
At most 4	0.294102	16.22723	18.39771	0.0980
At most 5*	0.160690	5.430414	3.841466	0.0198
		Max-Eigen Statistic		
None*	0.815524	52.39731	43.41977	0.0041
At most 1*	0.699349	37.25597	37.16359	0.0488
At most 2*	0.631478	30.94590	30.81507	0.0482
At most 3	0.518533	22.65844	24.25202	0.0800
At most 4	0.294102	10.79682	17.14769	0.3279
At most 5*	0.160690	5.430414	3.841466	0.0198

Source: Calculated by author, *denotes rejection of the hypothesis at the 0.05 level, **MacKinnon et al. (1999) p-values

Table 12.4 Estimated VECM

Error Correction:	dy_t	dx_{1t}	dx_{2t}	dx_{3t}	dx_{4t}	dx_{5t}
CointEq1	−1.1423	0.0052	−0.00066	1.8085	−67.7526	26.14421
	[−5.049]*	[1.192]	[−0.932]	[1.648]	[−0.097]	[0.882]
CointEq2	43.4007	−1.2102	−0.0051	114.0461	−2663.023	5054.500
	[2.868]*	[−4.137]*	[−0.108]	[1.554]	[−0.057]	[2.550]*
CointEq3	−62.877	−2.2092	−1.0907	−586.69	213012.7	19319.51
	[−0.645]	[−1.172]	[−3.544]*	[−1.242]	[0.713]	[1.513]**
dy_{t-1}	0.2044	−0.0105	0.0003	−1.6225	−26.775	30.169
	[1.272]	[−3.400]*	[0.639]	[−2.083]*	[−0.054]	[1.433]**
dx_{1t-1}	−16.793	−0.2364	0.0201	−45.066	9615.81	−1890.917
	[−1.904]*	[−1.386]**	[0.723]	[−1.053]	[0.355]	[−1.636]**
dx_{2t-1}	−0.4281	1.7420	0.1284	330.331	−19012.39	−20937.27
	[−0.0061]	[1.290]	[0.582]	[0.976]	[−0.088]	[−2.289]*
dx_{3t-1}	−0.0260	−0.0013	0.0001	0.0740	−180.59	12.718
	[−0.516]	[−1.369]**	[0.871]	[0.302]	[−1.167]	[1.921]*
dx_{4t-1}	$-6.62e^{-05}$	$-1.48e^{-06}$	$-2.33e^{-07}$	0.00014	−0.46330	−0.00108
	[−0.782]	[−0.902]	[−0.873]	[0.355]	[−1.787]*	[−0.098]
dx_{5t-1}	−0.000804	$-1.74e^{-05}$	$-7.04e^{-06}$	0.01524	2.2425	−0.1152
	[−0.541]	[−0.604]	[−1.499]**	[2.114]*	[0.492]	[−0.591]
C	0.3804	−0.0144	−0.0031	−6.5150	12828.57	1595.729
	[0.029]	[−0.058]	[−0.077]	[−0.104]	[0.326]	[0.948]
@TREND(90)	0.0572	−0.0005	$-1.02e^{-05}$	0.3162	−724.81	−90.574
	[0.089]	[−0.045]	[−0.005]	[0.101]	[−0.369]	[−1.078]
R-squared	0.6517	0.8633	0.5635	0.6807	0.2939	0.5763
F-statistic	3.7422	12.631	2.5825	4.2646	0.8324	2.7211
Akaike AIC	9.9537	2.0620	−1.5615	13.111	26.008	19.704
Schwarz SC	10.462	2.5709	−1.0526	13.619	26.516	20.212

Source: Calculated by author, * = significant at 5% level, ** = significant at 10%, d = first difference

Table 12.5 Values of roots

Roots	modulus
1.000000	1.000000
1.000000	1.000000
1.000000	1.000000
−0.804716	0.804716
−0.238034 – 0.689079i	0.729033
−0.238034 + 0.689079i	0.729033
0.037016 – 0.686182i	0.687180
0.037016 + 0.686182i	0.687180
−0.528179 – 0.270202i	0.593281
−0.528179 + 0.270202i	0.593281
−0.156071 – 0.486457i	0.510880
−0.156071 + 0.486457i	0.510880

Source: Calculated by author

The estimated equations of VECM implied that the change of temperature change and change of average rainfall are negatively related with the change of food supply in kilo calorie per capita per day in India significantly during 1990–2023. The change of temperature change and the change of co_2 emission is positively related significantly. The change of methane emission and change of food supply in kilo calorie per capita per day is positively related significantly. The change of rainfall is positively related with change in methane emission. The change in methane emission is negatively related with change in emission from agricultural land use, and positively related with change in average rainfall respectively. All other relations are statistically insignificant.

The VECM consists of 3-unit roots and 8 imaginary roots which are less than one and the remaining 1 root is negative so that the model is nonstationary. In Table 12.5, The values of roots are shown.

All the roots lie on or inside the unit circle for which the VECM is stable which is seen the Fig. 12.1.

The response of change in food supply in kilo calorie per capita per day (dy) to change in emission from agricultural land use (dx_2) reached the equilibrium after 2.5, 5, 6.5 and 9 years then departed away. The response of dy to change in average rainfall (dx_3) reached equilibrium after 2.5, 4, 6, 7, and 9.5 years then converged to equilibrium. The response of dx_1 to dy reached equilibrium at 1, 2.5, and 3.5 years then departed away. The response of dx_2 to dy reached at 1.5, 3, 4, 5.5, 8, and 9 years then diverged. It is noted that responses of increment in temperature change (dx_1) to dx_2, change in CO_2 emission (dx_4) and change in methane emission (dx_5) reached equilibrium many times but finally moved away. The responses of dx_3 to dx_4 and dx_5 reached equilibrium many times but diverged finally. The response of dx_4 to dx_5 reached equilibrium many times and moved away finally. All the responses in VECM are shown in Figs. 12.2a and 12.2b.

There are significant short run causalities from the changes of food supply in kilo calorie per capita per day to the increment of temperature change and average rainfall respectively. The short run causality from changes in methane emission to changes in emission in agricultural land use is significant. Even, the short run

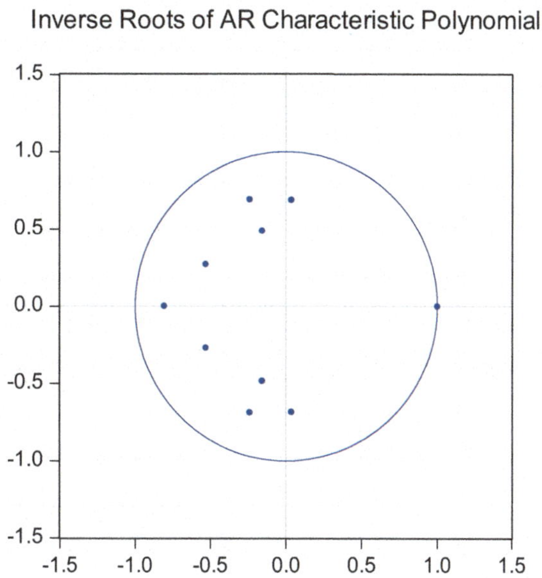

Fig. 12.1 The unit circle. Source: Plotted by author

causality from changes in average rainfall to changes in methane emission is also significant. From the system equations, the Wald test (1943) showed that the Chi-square values are rejected from null hypothesis of no causality. All the values have been shown in Table 12.6.

The long run associations or long run causalities were obtained from the significant three cointegrating equations which were found from the VECM. These cointegrating equations are given below.

$$Z_{1t-1} = -1.1423dy_{t-1} - 0.2664dx_{3t-1} - 8.93e^{-05}dx_{4t-1} - 0.005217dx_{5t-1} - 0.46107t + 38.798$$
$$(-5.049)* \quad (-4.088)* \quad (-1.166) \quad (-4.321)*$$

$$Z_{2t-1} = -1.21023dx_{1t-1} - 0.00392dx_{3t-1} + 1.17e^{-06}dx_{4t-1} - 5.13e^{-05}dx_{5t-1} + 0.0024t + 0.207$$
$$(-4.13)* \quad (-3.42)* \quad (0.868) \quad (-2.41)*$$

$$Z_{3t-1} = -1.0907dx_{2t-1} + 0.00048dx_{3t-1} + 3.05e^{-08}dx_{4t-1} - 6.26e^{-06}dx_{5t-1} - 0.000605t - 0.0169$$
$$(-3.54)* \quad (3.03)* \quad (0.160) \quad (-2.09)*$$

$$(* = \text{significant at } 5\% \text{ level, } t = \text{trend, } Z = \text{normalised variable}).$$

The first cointegrating equation states that the increment in food supply in kilo calorie per capita per day is negatively associated with the changes of average rainfall in mm, changes in CO_2 emission in kilo ton (insignificant at 5% level) and changes in methane emission in kilo ton in India during 1990–2023 significantly. The cointegrating equation has been significantly converging towards the equilibrium point at the speed of adjustment 114.23% per year. It is converging because the

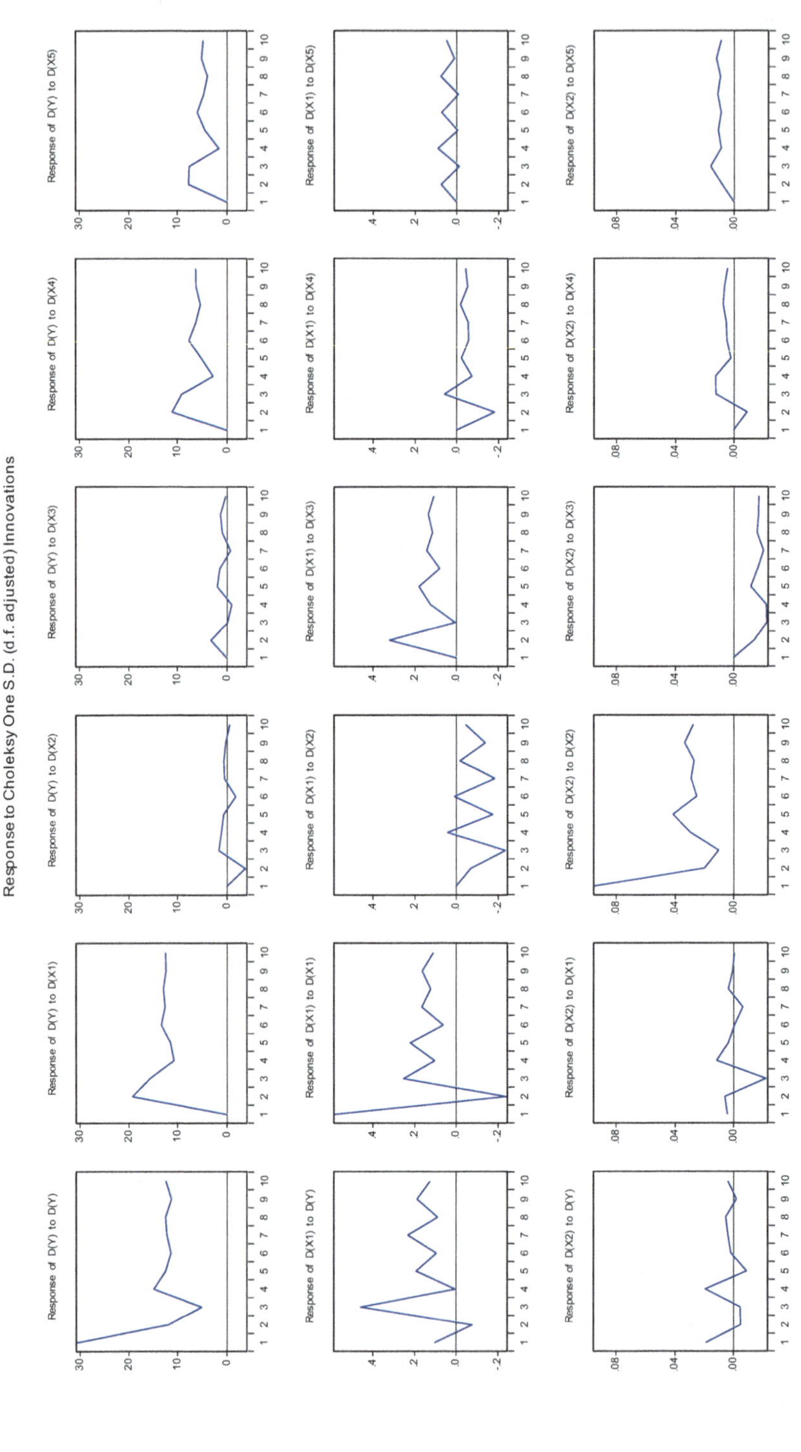

Fig. 12.2a Impulse response functions. Source: Plotted by author

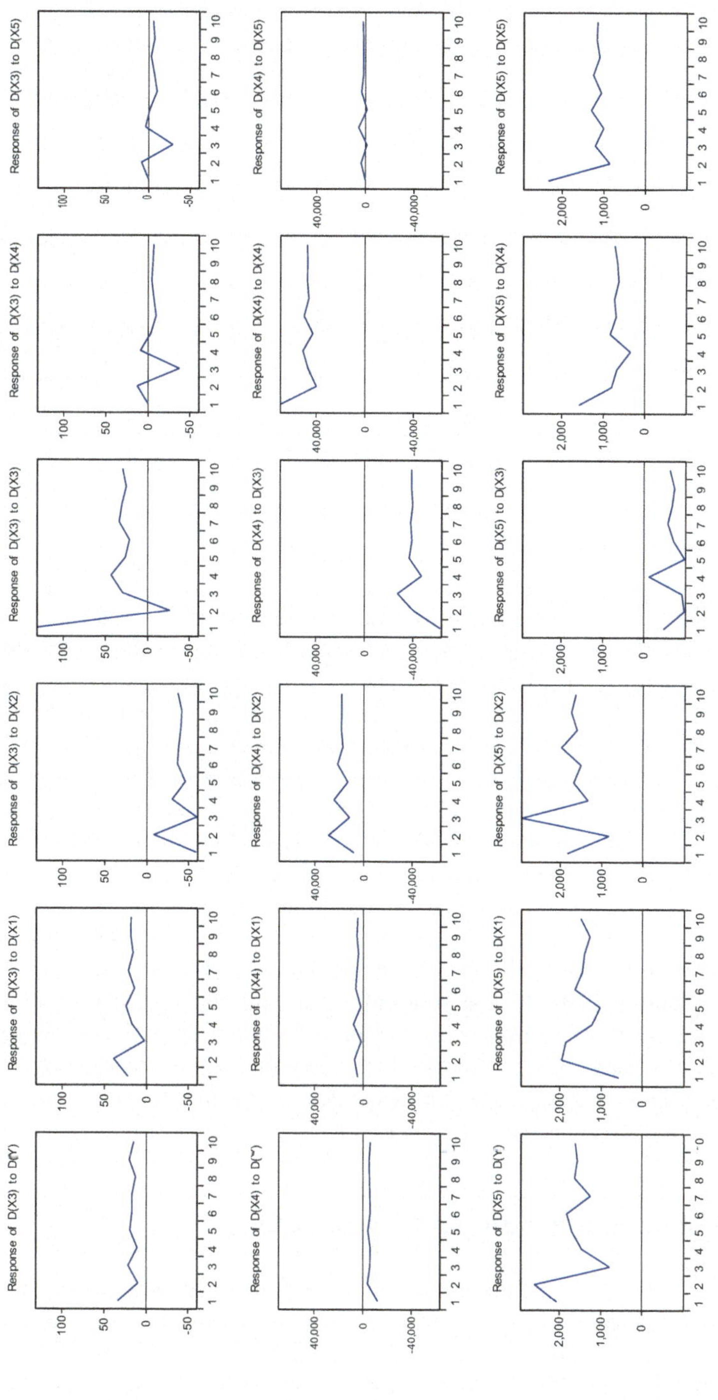

Fig. 12.2b Impulse response functions. Source: Plotted by author

Table 12.6 Short run causalies

Short run Causality from … to …	Chi-square (1)	Probability	H0 = no causality
From dy_{t-1} to dx_{1t}	11.560	0.0007	Rejected
From dy_{t-1} to dx_{3t}	4.341	0.037	Rejected
From dx_{5t} to dx_{3t}	4.473	0.034	Rejected
From dx_{2t} to dx_{5t}	5.243	0.022	Rejected

Source: Calculated by author

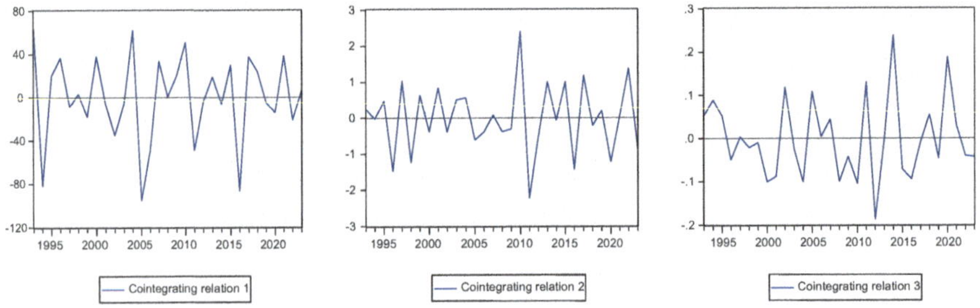

Fig. 12.3 Cointegrating equations. Source: Plotted by author

coefficient of dy_{t-1} is negative and its t value is significant at 5% level. Its trend value of the coefficient is negative. The second cointegrating equation states that the increment in temperature change is negatively associated with the changes of average rainfall in mm, and changes in methane emission in kilo ton and positively related with changes in CO_2 emission(insignificant) in India during 1990–2023 significantly. The second cointegrating equation has been approaching toward equilibrium at the speed of adjustment 121.02% per year. It is converging because the coefficient of dx_{1t-1} is negative and its t value is significant at 5% level. The trend coefficient is positive. The third cointegrating equation states that the increment in emission from agricultural land use in ton per hectares is positively associated with the changes of average rainfall in mm and changes in CO_2 emission in kilo ton (insignificant) and negatively associated with changes in methane emission in kilo ton in India during 1990–2023 significantly. The third cointegrating equation has been converging toward equilibrium at the speed of adjustment 109.07% per year. It is converging because the coefficient of dx_{2t-1} is negative and its t value is significant at 5% level. The trend coefficient is negative. All the significant converging cointegrating equations have been depicted in Fig. 12.3.

12.6 Policy Suggestions and Discussions

As regards the results of the present model is concerned, it claims to [i] reduce the emissions of agricultural land use, [ii] to decrease the emissions from green-house gases, [iii] to reduce the emissions from the methane gas emissions which can improve the supply of food in kilo calorie per capita per day in India. Moreover, the model claims that the reductions of increment of temperature change and the increment of average rainfall in India to counteract the food insecurity should be initiated.

All the policies primarily encourage to rise the agricultural food production by adopting crop rotations, crop diversification, using more mechanisation, irrigation and area of cultivation, applying scientific methods of water management and research, applying GIS in agriculture and so on. On the other hand, procurement of food and proper distribution of food is urgently needed. To protect climate change indicators, adaptation and mitigation techniques should be implemented. Paris agreement for protection of rising emission should be implemented through stages of plans. All the disaster management against climate change impacts must be reserved on behalf of the government.

Most important policy concern is that to meet future consumption trends, food production should be increased by 60% within 2050 which will also increase GHG emissions from agriculture where IPCC reminds to reduce emission within 2100 to limit global warming by 2 °C. Therefore, the countries must reduce emission to meet the future goal of food supply (Alexandratos & Bruinsma, 2012). The crucial and fundamental challenge is to address adaptation and mitigation policies together in the context of food security, at farm, national and global levels (Campbell et al., 2016). Output based intensity approach can fulfil both food security and climate change mitigation in reduction of agricultural emissions with higher yield (Murray & Baker, 2011).

The most innovative approach is climate-positive food system transformation which demands soft infrastructure such as digital climate service, innovative insurance tools, advisory services and actionable information, and financial services to support increased productivity and sustainability in both rural and urban food systems (Swinnen et al., 2022).

12.7 Limitations and Future Scope of Research

The long run data for more than 50 years are important to relate climate change indicators and food security indicators for obtaining meaningful and good results. The indicators like emission intensity from rice, cereals and other agricultural crops, nutritional food supply should be taken for considerations for better results. Therefore, better study can be done in future.

12.8 Conclusions

The paper concludes that food supply in kilo calorie per capita per day in India from 1990 to 2023 has long run association with temperature change, emission from agricultural land use, average rainfall, emissions from CO_2 and methane where there are 4 significant cointegrating equations in Trace statistic and 3 significant cointegrating equations in Max Eigen statistic as shown by Johansen cointegration rank test of those first difference series. The vector error correction states that the increment of temperature change and change in average rainfall are negatively related significantly with increment of food supply in kilo calorie per capita per day while increment of emission from methane has positively related with increment of food supply in kilo calorie per capita per day. There are significant short run causalities of increment of food supply in kilo calorie per capita per day from increment of temperature change and from increment of rainfall. Besides, there are significant long run causalities of increment of food supply in kilo calorie per capita per day from increment of average rainfall, increment of emissions from CO_2 and methane respectively where relationships are negative. All cointegrating equations have been converging towards equilibrium significantly. The impulse response of change in food supply in kilo calorie per capita per day to change in emission from agricultural land use reached equilibrium after 2.5, 5, 6.5 and 9 years respectively and then departed away.

Acknowledgement I am very much thankful to Lincoln University College, Malaysia and the authorities of references. I am responsible for all errors and omissions.

Sources of Funds No government or NGOs or educational institutions have sanctioned any funds for completion of this research.

Conflicts of Interest There is no conflict of interest in the publication of the paper.

References

Affoah, R., Zheng, H., Dengui, K., & Dissani, B. M. (2022). The impact of climate variability and change on food security in sub-Saharan Africa: Perspective from panel data analysis. *Sustainability, 14*(2), 759. https://doi.org/10.3390/su14020759

Alexandratos, N., & Bruinsma, J. (2012). *World agriculture towards 2030/2050: The 2012 revision*. (ESA Working Paper, 12-03). FAO Agricultural Development Economics Division. http://www.fao.org/docrep/016/ap106e/ap106e.pdf.

Bansal, V. (2022, July 11). *Food insecurity and hunger continue to plague India*. https://www.deccanherald.com/business/food-insecurity-and-hunger-continue-to-plague-india-1125469.html.

Bhat, S. U., & Pandit, A. K. (2020). Water quality assessment and monitoring of Kashmir Himalayan freshwater springs—A case study. *Aquatic Ecosystem Health & Management, 23*(3), 274–287. https://bioone.org/journalArticle/Download?fullDOI=10.1080%2F14634988.2020.1816771.

Bhuyan, B., Mohanty, R. K., & Patra, S. (2023). Impact of climate change on food security in India: An evidence from autoregressive distributed lag model. *Environment, Development and Sustainability, 27*, 6349–6369. https://doi.org/10.1007/s10668-023-04139-3

Campbell, B. M., Vermeulen, S. J., Aggarwal, P. K., Corner-Dolloff, C., Girvetz, E., Loboguerrero, A. M., Ramirez-Villegas, J., Rosenstock, T., Sebastian, L., Thornton, P. K., & Wollenberg, E. (2016). Reducing risks to food security from climate change. *Global Food Security, 11*, 34–43. https://doi.org/10.1016/j.gfs.2016.06.002

Dasgupta, S., & Robinson, E. J. Z. (2022). Attributing changes in food insecurity to a changing climate. *Scientific Reports, 12*, 4709. https://www.nature.com/articles/s41598-022-08696-x.

FAO. (2020). *The state of food security and nutrition in the world 2020. Transforming food systems for affordable gealthy diets*. FAO. Available online at: http://www.fao.org/3/ca9692en/ca9692en.pdf.

IPCC. (2019). *Climate change and land: An IPCC Special Report on climate change, desertification, land degradation, sustainable land management, food security, and greenhouse gas fluxes in terrestrial ecosystems*. https://www.ipcc.ch/srccl/.

Jeder, H., Hattab, S., & Frija, I. (2020). An econometric analysis for food security in Tunisia. *New-Medit, 19*(4–01) https://newmedit.iamb.it/bup/wp-content/uploads/2020/12/New-Medit-2020-Vol-XIX-n.4-01-An-econometric-analysis.pdf.

Johansen, S. (1988). Statistical analysis of cointegrating vectors. *Journal of Economic Dynamics and Control, 12*, 231–254. https://econpapers.repec.org/RePEc:eee:dyncon:v:12:y:1988:i:2-3:p:231-254.

Kaur, H., & Kaur, S. (2018). Climate change impact on agriculture and food security in India. *Journal of Business Thought, 7*, 35–62. https://www.informaticsjournals.com/index.php/jbt/article/view/21329.

Khan, A. A., & Hassan, A. (2017). Climate change: Concern for food security in India. *IOSR Journal of Humanities and Social Science, 22*(10), 52–57. https://www.iosrjournals.org/iosr-jhss/papers/Vol.%2022%20Issue10/Version-5/H2210055257.pdf.

Kumar, A., & Sharma, P. (2022, June). *Impact of climate variation on agricultural productivity and food security in rural India*. https://ssrn.com/abstract=4144089 or https://doi.org/10.2139/ssrn.4144089

MacKinnon, J., Haug, A., & Michelis, L. (1999). Numerical distribution functions of likelihood ratio test for cointegration. *Journal of Applied Econometrics, 14*(5), 563–577. https://qed.econ.queensu.ca:80/jae/1999-v14.5/.

Mbow, C., et al. (2019). Food security. Climate change and land: An IPCC special report on climate change, desertification, land degradation, sustainable land management, food security, and greenhouse gas fluxes in terrestrial ecosystems. In P. R. Shukla et al. (Eds.), *Intergovernmental panel on climate change*. https://www.ipcc.ch/site/assets/uploads/2019/11/SRCCL-Full-Report-Compiled-191128.pdf.

Murray, B. C., & Baker, J. S. (2011). An output-based intensity approach for crediting greenhouse gas mitigation in agriculture: Explanation and policy implications. *Greenhouse Gas Measurement & Management, 1*, 27–36. https://www.tandfonline.com/doi/epdf/10.3763/ghgmm.2010.0004?src=getftr.

Singh, N.P., Anand, B., & Singh, S. (2020). *Impact of climate change on agriculture in India: Assessment for agro-climatic zones (Policy Paper 35)*. ICAR-National Institute of Agricultural Economics and Policy Research (NIAP), https://niap.icar.gov.in/pdf/pp35.pdf.

Swinnen, J., Arndt, C., & Vos, A. R. (2022). Climate change and food systems transforming food systems for adaptation, mitigation, and resilience. In *2022 global food policy report: Climate change and food systems*. IFPRI. https://ebrary.ifpri.org/utils/getfile/collection/p15738coll2/id/135884/filename/136107.pdf.

Wald, A. (1943). Test of statistical hypothesis concerning several parameters when the number of observations is large. *Transactions of American Mathematical Society, 54*, 426–482. https://www.pp.rhul.ac.uk/~cowan/stat/wald1943.pdf.

A Study of Economic Viability of Millet Cultivation in India

13

Jyoti Kumari Mishra

Abstract

The chapter will provide in-depth analysis of the economic viability of millet cultivation in India which is providing deep insight into the assessment of input cost of millets and enlightening the nutritional benefits and challenges of millets cultivation in India. The chapter is also focusing on then study of economic feasibility of millets cultivation in India.

The study is based on secondary data which have been collected through different published sources of information like newspapers, journals, magazines and different government and non-government websites. The collected statistical data is presented through graphs and diagrams for analysis purpose.

Conclusion is aiming on the achievement of economic viability of millet cultivation in India which consist holistic solutions and combined efforts of policyholders and stakeholders. By addressing the related challenges, grasping opportunities, and adopting sustainable approaches, we can unleash the millets' economic potential, improve farmer (Stakeholder') livelihoods and socio-economic status. The conclusion is also providing in depth insight about how food security and nutritional values of millets can be enhanced. This is also providing transforming India's agricultural landscape for present and future generations.

Keywords

Millet cultivation · Economic viability · Input cost of millet cultivation · Determinants of millet cultivation

J. K. Mishra (✉)
Department of Banking Economics and Finance, Bundelkhand University, Jhansi, UP, India

C. B. Singh et al. (eds.), *Global Millets Production for a Sustainable Future*,
https://doi.org/10.1007/978-3-031-88157-2_13

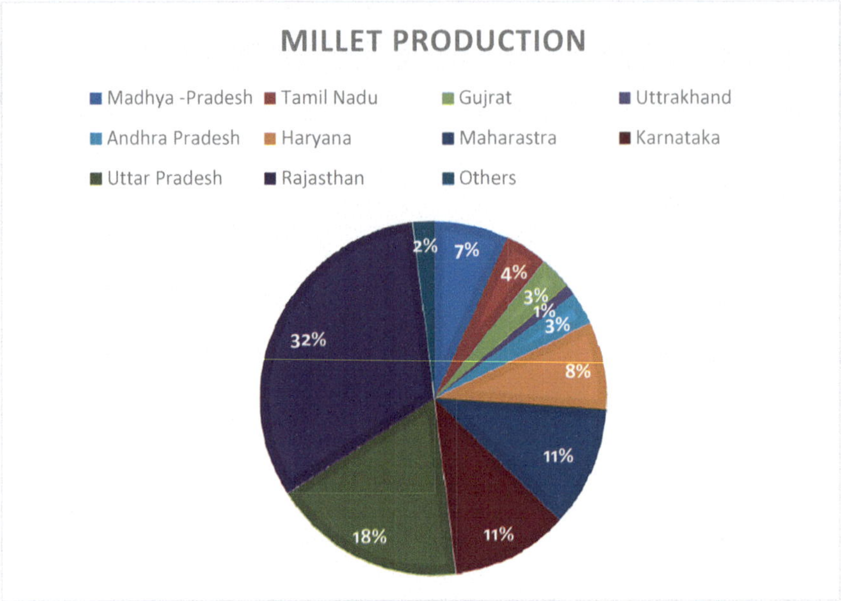

Fig. 13.1 State wise Millet Production 2023–2024 (2n advance estimates). Source: Agricultural and processed food products export development authority (APEDA)

13.1 Overview About Millet

Millets are among the earliest food that people have ever eaten and may have been the first cereal grain to be utilized in a home. Millets are tough, small-seeded grasses that thrive in arid regions as rain-fed crops with mediocre soil fertility and moisture levels. Millet is a collective term for several distinct small-grained cereal grasses. Despite not coming from a single plant species, millet shares a number of constant characteristics Shanmugam (2020). The word "Millet" derived from the latin word "Milium" means small seed Morya et al. (2017). No doubt India is the biggest producer of Millets in the world. India contributes almost 41% of yield of Millet in comparison to those countries who are contributing enough in millet production for this sake and to establish India in Millet growing countries Government of India Declared 2023 'International Year of the Millet'. In India major producer of Millets are Rajasthan Haryana Gujrat Madhya Pradesh, Tamil Nadu, Andhra Pradesh, Maharashtra, Uttar Pradesh and Uttarakhand (Karuppasamy, 2015). The projected yield of Millets in 2023 and in 2024 is depicted by the following Pie chart (Fig. 13.1).

Pie chart is revealing that the highest percentage of millet Yield is in Uttar Pradesh and the lowest percentage is Uttarakhand as per the estimates of 2023 and 2024. this is revealing that Uttar Pradesh have big potential for millet production and marketing. Millet is produced in many varieties. According To APEDA the production of following varieties is estimated to be produced (Fig. 13.2).

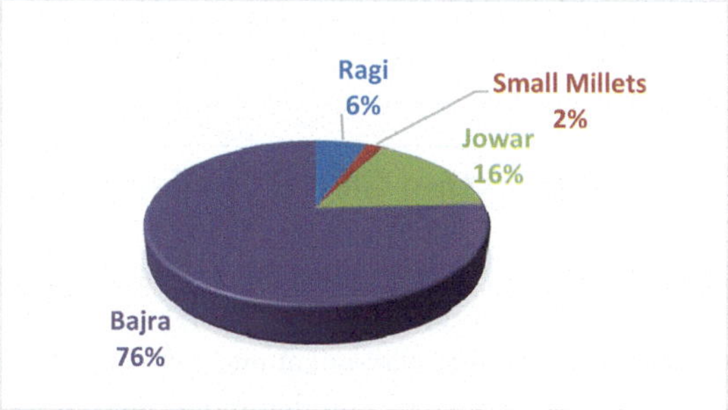

Fig. 13.2 Production % of different millets according to APEDA. Source: Agricultural and processed food products export development authority (APEDA)

13.2 Cultivation Practices

India's millet production is characterized by the quantity and relative importance of the various varieties, their geographic dispersion, ecological adaptability, nutritional significance, reaction to input and management methods, processing and value addition, and research and policy interventions Anil Shahane & Singh Shivay (2023). The cultivation of Millet varies on the specific variety and environmental conditions. Generally, millets are grown in the rainy season and harvested after maturation of the crop. Traditional farming methods such as intercropping (bajra) Balasubramanian et al. (1984) and mixed cropping are commonly used to maximize yield and enhance soil fertility. Millets are often cultivated by marginal or small holding farmers using sustainable practices that minimizes the use of chemical inputs and promotes biodiversity, and they can be also grown in diverse agro-ecological conditions (Klmata et al., 2000).

13.3 Economic Importance

Millets are providing the livelihoods of millions of people, particularly in developing countries where millets are served as staple food crops. They contribute to food security and income generation for smallholding farmers and providing a strong source of livelihood even in challenging agro-climatic conditions. Apart from it, millets are having economic potential with opportunities for value addition and commercialization. Millet-based products such as flour, breakfast cereals, snacks, and beverages are gaining popularity day by day in urban and even in rural markets also, Millet cultivation is creating new market opportunities for farmers and entrepreneurs.

13.4 Role of Millets in Sustainable Agriculture

Millets does not rely on the utilisation of the synthetic fertilisers they can be grown by bio fertilisers and fertilisers of waste. Less pesticides are also used in the cultivation of millets. so, when millets are cultivated with traditional method and use of less pesticides provides resistance against pesticides and are called pesticide free crops. So, they are significantly contributing to the agriculture eco system.

13.5 Addressing Global Challenges

Millets have all the attributes and potential to address numerous pressing global issues like food security, malnutrition, and climate changing conditions. As Millet preserves soil health, ecological integrity and soil vitality. On August 3/2024 prime minister Narendra Modi said that India can find solution in the form of millet cultivation to address global challenges on the nutritional value, water scarcity and climate change and can develop sustainable agriculture with good food systems and can tackle the holistic approach of one Earth, One family and One future.

13.6 Economic Viability of Millet Cultivation

Assessing the economic viability of millet cultivation is of paramount importance for farmers, policymakers, and stakeholders for economic viability can be assessed on the basis of time span in short term, viability can be assessed on the basis where other crops have less importance because there are no favourable conditions to grow them but millets can be while in long term, millet viability can be increased by supporting government policies. Millets can be said economic viable also because millets can be cultivated in lower input cost and also can be cultivated in drought resistance areas.

13.7 Importance of Study

The study on the economic viability of millet cultivation in India holds significant importance on various fronts. Firstly, millets are traditional grains that have been cultivated for centuries in India but have faced neglected in recent times due to the emphasis on high-yielding crops like rice and wheat. Understanding the economic viability of millet cultivation can help in reviving these indigenous crops, promoting agricultural diversity, and ensuring food security. Moreover, millets are highly resilient to climatic variations and require fewer inputs such as water and fertilizers compared to other cereals, making them suitable for sustainable agriculture. The study provides in depth understanding of millets nutritional values.

Objectives of the Study
The objectives of the study are following:

1. To know the nutritional benefits of millets
2. To know the challenges of millets cultivation.
3. To study the economic feasibility of millets cultivation in India.

Table 13.1 Types of millets and their nutritional source

S. No.	Variety of millet	Source of vitamin	Source of minerals	Preventive disease
1.	Finger millet (Ragi)	Vitamin A vitamin B and niacin	Rich in amino acids and antioxidant	Helpful in brain growth
2.	Foxtail millet (Kakun/Kangni)	Vitamin B 12	Carbohydrates, iron	Improves immunity, and good for nervous system
3.	Sorghum millet (Jowar)	Vitamin B1, Vitamin B6	Iron, protean and fiber	Increases metabolism
4.	Pearl millet (Bajra)	Vitamin A	Iron, fibre, protein, magnesium and calcium	Protects from type II diabetes
5.	Buckwheat millet (Kuttu)	Vitamin B6	Magnesium, phosphorous	Protects from cancer, asthma in children and gallstones
6.	Amarnath millet (Rajgir/Ramdana/Chola)	Vitamin C	Protein and dietary fibre	Helpful in fighting greying hair and hair loss
7.	Little Millet (Moraiyo/Kutki/Shavan/Sama)	Vitamin B	Calcium, iron, zinc and potassium	Helpful in weight loss
8.	Kodo millet (Barnyard millet	Vitamin B6	Dietary fibre, calcium, folic acid and phosphorus	Helpful in weight loss and strengthens bone density
9.	Broomcorn millet	Vitamin E, vitamin B1, B2, B6	Zinc, copper and minerals	Maintains blood sugar level
10.	Kodo millet	Vitamin B	Calcium, iron, potassium, and zinc	Checks blood pressure and cholesterol level

Source: Jacob et al. (2024)

13.8 Nutritional Components of Millets

Millets are highly nutritious grains which offers a range of essential nutrients such as protein, dietary fibre, vitamins, anti-oxidants, minerals and are non-glutinous and non-acid forming. Millets are gluten-free, which makes them suitable for individuals and protects them from celiac disease or gluten intolerance (Jacob et al., 2024; Umanath 2018; Malaiarasan et al. 2022). Millets have a low glycaemic index, which makes ideal for managing blood sugar levels, specifically for those individuals who are suffering from diabetes. A variety of millets like Sorghum, pearl millet, finger millet, foxtail millet, and proso millet are the most commonly consumed millets each variety offers unique nutritional benefit. The types of millets and their nutritional content can be also presented with the help of Table 13.1.

Table 13.1 and all content of this chapter is providing an eye view on the importance of millet that how millet can be a good substitute and component of super and healthy food. This can be cultivated in low cost and can provide high nutritional values. Now Government is more emphasing on its cultivation and also focusing its concentration on products made with it. Its market share is also expanding day by day. No doubt that Millet is a future food and coming back as it was an important part of our Thali in olden days. Government is also focusing that millet should be contributing grain in food security. Millets have gained recognition for their resilience to harsh environmental conditions, nutritional benefits, and potential contribution to food security and sustainable agriculture.

13.9 Challenges in Millet Cultivation in India

Millet is important food component and government is promoting its cultivation at large scale. Apart from it there are many difficulties in the cultivation of millets.

- Lack of perfect knowledge about the economic feasibility and profitability farmers are not aware about the modern methods of cultivation so they get lower productivity from millet cultivation.
- Imperfect knowledge about market conditions and they are nit familiar with the methods how market demand can be created and increased.
- Farmers do not have direct connectivity with the consumers or manufacturers for millet products where they can directly supply the raw material and can get proper price of their produce García-Granero et al. (2016).

13.10 Economic Feasibility of Millet Cultivation

Assessing the economic viability of millet cultivation is of paramount importance for farmers, policymakers, and stakeholders for economic viability can be assessed on the basis of time span in short term, viability can be assessed on the basis where other crops have less importance because there are no favourable conditions to grow them but millets can be while in long term, millet viability can be increased by supporting government policies (Prasad Sati, 2023). A perfect strategy regarding marketing and branding should be implemented to promote millet products & Minimum support price should be also announced by the government. Millets can be said economic viable also because millets can be cultivated in lower input cost and also can be cultivated in drought resistance areas.

13.11 Conclusion

The study of economic viability of millet cultivation in India presents a promising and leading trajectory for agricultural sustainability and better eco-system with rural development. A comprehensive study is projected to delve into the comparative analysis of millet cultivation against traditional crops with the assessing factors like input costs, market demand, and resilience to changing climatic conditions. The study will provide that how farmers and policymakers can align themselves with the potential profitability and feasibility of millet cultivation with the existing agricultural practices. And how pre decided interruptions can enhance the value chain of millet production. The cultivation and promotion of millet in India have many potential pathways for future development. The future development can be accessed by practicing some valuable suggestions like promotion and awareness about millet cultivation and marketing with research and development. There should be international collaboration and integration of food system overall a comprehensive approach is required to promote millets in India to promote rural development and upliftment of farmers.

References

Anil Shahane, A., & Singh Shivay, Y. (2023, July 1). Millets cultivation in non-conventional zone- practices, possibilities and prospects. *Acta Scientific Agriculture, 7*(7), 3–12. https://doi.org/10.31080/asag.2023.07.1274

Balasubramanian, T. N. J. G., et al. (1984). Study on the impact of intercropping in Bajra. *Madras Agricultural Journal, 71*(11), 705–708. https://doi.org/10.29321/maj.10.a02538

García-Granero, J. J., Lancelotti, C., Madella, M., & Ajith Prasad, P. (2016). Millets and herders: The origins of plant cultivation in semiarid North Gujarat (India). *Current Anthropology, 57*(2), 149–173. https://doi.org/10.1086/685775

https://www.assocham.org/uploads/files/Report_Millet

https://www.google.com/search?q=area+production+and+productivity+of+millets+in+world&source=lnms&tbm=isch&sa=X&ved=2ahUKEwj6r9uvo7L9AhVz_3MBHcPZAfIQ_AUoAXoECAEQAw&biw=731&bih=343&dpr=2.63

https://www.google.com/search?q=market+size+of+millets+and+its+products+in+the+market&source=lmns&bih=343&biw=731&hl=en-US&sa=X&ved=2ahUKEwjvi6-42LL9AhWM-jgGHTJuDDsQ_AUoAHoECAEQAA

https://www.google.com/search?q=millets&tbm=isch&chips=q:millets,g_1:grain:I3jFZ477M2c%3D&hl=enUS&sa=X&ved=2ahUKEwiwqrG7trL9AhXf4nMBHawUAAYQ4lYoBnoECAEQMA&biw=722&bih=334

https://www.maximizemarketresearch.com/market-report/global-millet-market/105674/

https://www.researchgate.net/figure/State-wise-area-production-and-yield-of-pearl-millet-in-India_tbl1_349313137

Jacob, J., Krishnan, V., Antony, C., Bhavyasri, M., Aruna, C., Mishra, K., Nepolean, T., Satyavathi, C. T., & Visarada, K. B. R. S. (2024). The nutrition and therapeutic potential of millets: An updated narrative review. *Frontiers in Nutrition, 11*. https://doi.org/10.3389/fnut.2024.1346869

Karuppasamy, P. (2015). Overview on millets. *Trends in Biosciences, 8*(13), 3269–3273.

Klmata, M., Ashok, E. G., & Seetharam, A. (2000, April). Domestication, cultivation and utilization of two small millets, brachiaria ramosa and Setaria glauca (Poaceae), in South India. *Economic Botany, 54*(2), 217–227. https://doi.org/10.1007/bf02907825

Malaiarasan, U., Paramasivam, R., & Saravanakumar, V. (2022). Choice of millets cultivation in India: Evidence from farm household survey data. *Agricultural Economics Research Review, 35*(conf), 145–163. https://doi.org/10.5958/0974-0279.2022.00027.1

Morya, G. C. K., et al. (2017). Millets: The indigenous food grains. *International Journal of Advanced Ayurveda, Yoga, Unani, Siddha and Homeopathy, 6*(1), 447–452. https://doi.org/10.23953/cloud.ijaayush.328

Prasad Sati, V. (2023). Economic viability and prospects of oil palm cultivation in mizoram, India. *Tropical Agrobiodiversity, 4*(2), 62–67. https://doi.org/10.26480/trab.02.2023.62.67

Shanmugam, A. (2020, September 19). Millets cultivation area and quantity of consumption lead to doubling farmers income. *Adalya Journal, 9*(9). https://doi.org/10.37896/aj9.9/044

Umanath, M. (2018). Millets' consumption probability and demand in India: An application of Heckman sample selection model. *Economic Affairs, 63*(4). https://doi.org/10.30954/0424-2513.4.2018.29

Sorghum Production in Bundelkhand Region of Uttar Pradesh: A Comprehensive District Level Data Analysis from 2001 to 2022

14

Ankita Jasmine Lall, C. B. Singh, and Shilpa Mishra

Abstract

Aim: The Bundelkhand Region of U.P. has always been into picture due to its adverse agricultural climatic conditions and has especially been considered as a drought prone region. The climate conditions here support the production of millets but still over the years there is a decline in the production of this climate-smart agricultural crop.

Data: This study deals with time series data ranging from 2001 to 2022 based on the area, production and productivity of sorghum in Bundelkhand Region of U.P.

Objectives: The study aims to assess the trend on sorghum production in Bundelkhand Region of U.P.

Methodology: To assess the trend of sorghum production in the region Compound Annual Growth Rate was worked on.

Results: The analysis of sorghum production in the Bundelkhand region of Uttar Pradesh from 2001 to 2022 reveals several key trends. The total production of sorghum has seen a slight decline, with a Compound Annual Growth Rate (CAGR) of −0.96%. Banda showed the highest total contribution and a positive CAGR of 2.28%. The area under sorghum cultivation has also decreased significantly, with a CAGR of −3.42%. But slight stabilizations are seen in some years post-2013. Despite the reduction in cultivation area, the productivity of sorghum has increased, achieving a CAGR of 2.53%.

Conclusion: Sorghum production in Bundelkhand Region has increased in productivity despite a decline in cultivation area, making it crucial for food

A. J. Lall
Department of Economics, Christ Church College, Kanpur, Uttar Pradesh, India

C. B. Singh · S. Mishra (✉)
Institute of Economics and Finance, Bundelkhand University, Jhansi, Uttar Pradesh, India
e-mail: dr.cbsingh195@bujhansi.ac.in

C. B. Singh et al. (eds.), *Global Millets Production for a Sustainable Future*,
https://doi.org/10.1007/978-3-031-88157-2_14

security and can be a major contributor for reducing food inflation in India, especially in light of the International Year of Millets 2023. Promoting sorghum can enhance sustainable agriculture and economic stability in the region.

Keyword

Millet · Sorghum · CAGR · Sustainable development · Bundelkhand · Super Food

14.1 Introduction

With the growing global population, the demand for food will also rise. The global economies face severe food inflation resulting from various factors like supply chain disruptions, geo-political constraints, adverse weather conditions and rising input costs too. The ecological imbalances and extreme weather conditions have caused droughts and floods that have affected the crop yield further exacerbating the food supply.

The Malthusian Theory of Population can be linked to the current food inflation scenario. The theory justifies how the population grows at a geometric rate and food supply expands at an arithmetic rate leading to demand and supply gap where the population outpaces the food supply resulting in shortage in food supply and increase in food inflation. Though modern technologies have increased the production of food but still matching it with the challenges of present population growth and environmental imbalances have strained out the food supply and caused inflationary pressures.

The Sustainability Livelihood Framework accentuates the importance of diversifying agricultural practices to enhance resilience and sustainability. Millets with its ability to bear harsh climatic conditions proves to be climate-resilient that can align well with these conditions where other crops can miss the mark.

Even the agroecological theory can better support millet cultivation for a food security, linking the sustainable farming practices with natural ecosystem (Pandey et al., 2023). Millets require fewer inputs making them environmentally sustainable choice in production. The millets production is economically significant in regions facing water resource scarcity which results to be a major reason for millet production in India (Karthick et al., 2023). Also, the farmers can introduce millet production in their agricultural system and reduce their vulnerability to climate change and ensure a steadier food supply.

Millets despite being considered as a *'Super Food'* with immense nutritional benefits has confronted less consideration. Millet's production has the potential to withstand arid climatic conditions and require minimal resources thus ensuring to be climate-smart agriculture (Karthick et al., 2023) but still its production has been limited due to the consumption or market demand constraint.

This study is limited to sorghum production which is one of the main millets produced not only in India but at global level too. But amid this we can actually see that not only the production but also the consumption of sorghum at global level has remained constant. The global statistics on sorghum production reached its low in early 1990s but again after 2018 it is showing a rise. But a shift has been seen in the production trends moving away from Asia and America to Sub Saharan Africa (Dalton & Hodjo, 2021). Though India's global rank (third) in sorghum production lies after U.S. and Nigeria but still it is a significant contributor of the crop. Also, in India Maharashtra and Karnataka are the major Sorghum producing states. But this study considers the millet cultivation potential of Bundelkhand Region and tries to portray the trend of sorghum production in the region over a period of 20 years.

14.2 About Bundelkhand Region of U.P

The Bundelkhand Region of Uttar Pradesh lies in the Central India region. It is specified by its semi-arid climate and is commonly known to be a drought prone area. The region embraces seven districts of Uttar Pradesh; namely Jhansi, Jalaun, Hamirpur, Banda, Mahoba, Lalitpur and Chitrakoot upholding total geographical area of the region is 29,418 sq. km.

The Bundelkhand Region is one of the most backward regions of U.P. facing severe infrastructural, social, climatic, financial and agrarian constraint but still supporting 80% of the livelihood through agriculture (Singh & Mishra, 2019).

14.3 Agricultural Status of Bundelkhand Region

Undoubtedly, rainfall stands to be the prime factor for determining crop production in a particular area but specifically for Bundelkhand region the climate uncertainties has always affected the agricultural production and productivity. The region accounts for agriculture being the prime source of livelihood but deeply faces climate variability (Prakash et al., 2024). Historically, it was observed that the Bundelkhand Region of UP and MP has always experienced a drought in every 16 years in the nineteenth century. The data also revealed that during the period ranging from 1968 to 2000 this duration increased three times. Also, this region faced 5 drought years in the period ranging 2000 to 2010 (RAI et al., 2012). The major crops grown in the entire Bundelkhand Region include wheat, rice, gram and soyabean that fails to withstand the regions agricultural potential resulting in significant yield gaps due to erratic climatic conditions and resource restraint (Sharma et al., 2020). In addition to this millets production has also been seen in the Region but the trend is not that significant due to constant decline in its cultivation resulting from demand constraint. But the potential of millet production can support agricultural pattern in the region.

14.4 Sorghum Production in Bundelkhand Region

As stated earlier, millet production has faced significant fluctuation in the region but still amongst the total millets production sorghum holds a substantial position.

For this study secondary data on sorghum has been collected at district level for all seven districts comprising Bundelkhand Region of U.P. (Jhansi, Jalaun, Banda, Hamirpur, Chitrakoot, Mahoba and Lalitpur) for a period ranging from 2001 to 2022. This data encompassed the production, area and productivity of sorghum in region under study.

To analyse the data CAGR has been used. The CAGR has been estimated using linear function on time series data of sorghum production, area and productivity in the entire Bundelkhand Region of U.P (Table 14.1).

The above table provides an overview of **sorghum production** in the Bundelkhand region of Uttar Pradesh from 2001 to 2022. Total production across the region has slightly declined over the years, with a **Compound Annual Growth Rate (CAGR) of – 0.96%**. Total production in 2001 was **118,809 tonnes**, while in 2022 it stood at **107,455 tonnes**, reflecting a marginal overall decrease despite variability. If we see **district-wise performance**, **Banda** stands as the best-performing district in terms of production, with the highest total contribution (**603,710 tonnes**) and a positive CAGR of **2.28%**. **Chitrakoot**: Has shown slight growth, with a positive CAGR of **0.64%**. **Lalitpur** and **Mahoba**: Experienced the sharpest declines, with CAGRs of **−16.22%** and **−4.64%**, respectively. **Jhansi, Jalaun, and Hamirpur**: Moderate negative growth rates, reflecting a gradual decline in production. If we see the **best Year**, 2004 recorded the highest production (**122,117 tonnes**), followed closely by 2021 (**121,530 tonnes**) and 2016 was the lowest-performing year, with just **15,003 tonnes**, likely due to severe adverse conditions.

While overall production remains below early 2000 levels, the years 2020–2022 showed a recovery trend, with notable performances in districts like **Banda**, **Jalaun**, and **Chitrakoot** (Fig. 14.1).

Analysis

- **Declining Cultivation Area**: The significant reduction in land under sorghum cultivation directly impacts production.
- **Productivity Gains**: Despite area reductions, improvements in yield per hectare have helped mitigate declines in production, especially in districts like **Banda** and **Chitrakoot**.
- **Climatic Challenges**: The Bundelkhand region frequently experiences droughts and water scarcity, contributing to the variability and declining trends in sorghum production.
- **Shift in Cropping Patterns**: Farmers are switching to other crops with higher profitability or better market demand (Table 14.2).

The above table shows the area under sorghum cultivation in the Bundelkhand region of Uttar Pradesh from 2001 to 2022. The above table represents the overall decline in cultivation area. The total area under sorghum cultivation has decreased significantly over the years, with a **Compound Annual Growth Rate (CAGR) of −3.42%**. In the year 2001, the total area was **141,561 hectares**, while in 2022, it dropped to **71,858 hectares** that shows a reduction by nearly 50%. The **Yearly Trends** shows a most significant decline occurred between **2001 and 2013**, where cultivation area fell from 141,561 hectares to 73,587 hectares. Post 2013, the decline has been slower and shows

Table 14.1 Agricultural production of sorghum in Bundelkhand Region of UP (in Tonnes)

Year	Jhansi	Jalaun	Banda	Chitrakoot	Hamirpur	Lalitpur	Mahoba	Total
2001	6982	14,288	33,663	24,182	29,619	3792	6283	118,809
2002	4308	21,188	24,314	16,956	37,705	4668	4844	113,983
2003	695	8107	20,380	10,167	27,317	2414	4122	73,202
2004	5817	14,200	33,487	19,519	33,446	9962	5686	122,117
2005	3934	12,114	24,553	19,385	20,061	4737	5027	89,811
2006	1963	8799	23,646	20,641	11,362	1235	2619	70,265
2007	9296	19,857	19,708	10,088	15,858	6552	4399	85,758
2008	1282	2634	12,798	7500	3552	514	1921	30,201
2009	8702	16,744	10,856	11,480	6433	5694	5108	65,017
2010	1981	10,501	16,703	14,272	8432	3670	2865	58,424
2011	4114	9899	27,863	12,975	16,551	1148	4145	76,695
2012	2867	5738	28,505	18,951	18,902	591	3216	78,770
2013	3003	13,737	35,736	22,351	23,853	83	3861	102,624
2014	837	8402	27,040	11,833	16,574	111	2137	66,934
2015	1148	5877	26,097	11,378	12,989	12	922	58,423
2016	180	4126	7981	1598	897	53	168	15,003
2017	3108	7175	21,835	21,439	10,790	517	3751	68,615
2018	1640	12,919	38,262	13,360	14,164	362	1552	82,259
2019	1363	5706	31,138	13,677	6895	387	3181	62,347
2020	787	7689	41,439	28,103	14,727	206	2114	95,065
2021	1390	9075	56,178	31,645	16,938	923	5381	121,530
2022	4326	15,194	41,528	28,766	14,393	391	2857	107,455
Total	**69,723**	**233,969**	**603,710**	**370,266**	**361,458**	**48,022**	**76,159**	**1,763,307**
Mean	**3169.23**	**10634.95**	**27441.36**	**16830.27**	**16429.91**	**2182.82**	**3461.77**	**80150.32**
Std	**2576.30**	**4943.90**	**10986.35**	**7271.43**	**9177.85**	**2634.97**	**1568.85**	**27312.73**
LOGEST	**0.94**	**0.98**	**1.02**	**1.01**	**0.96**	**0.84**	**0.95**	**0.99**
CAGR%	**−5.74**	**−2.37**	**2.28**	**0.64**	**−4.33**	**−16.22**	**−4.64**	**−0.96**

Source: 1. Directorate of Economics and Statistics, Dept. of Agriculture and Farmers Welfare, Ministry of Agriculture and Farmers Welfare, Govt. of India
2. CAGR based on Author's Calculation

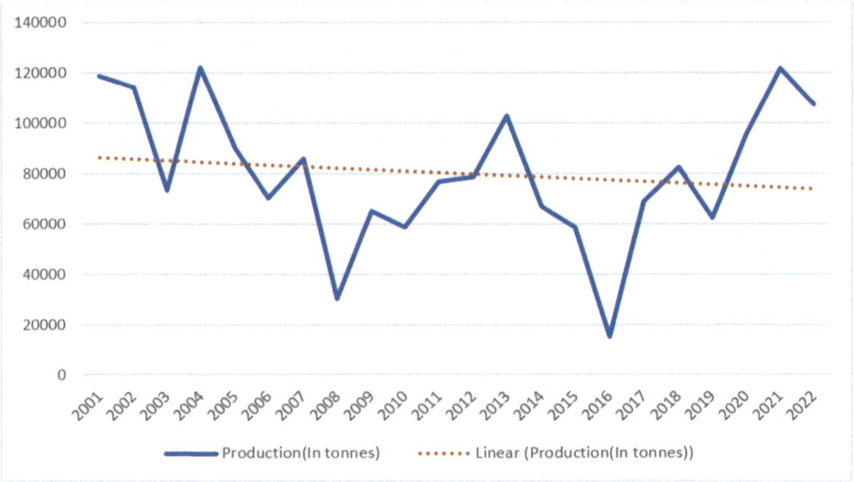

Fig. 14.1 Trend on production of sorghum in Bundelkhand Region of UP

slight stabilizations in some years (2020–2022). **Jhansi and Jalaun** consistently had the largest areas under cultivation but have also seen significant reductions in cultivated areas. **Chitrakoot**: While its area decreased, it remained relatively stable compared to other districts, particularly in the later years. **Lalitpur**: This district shows the sharpest decline, with areas nearly disappearing in certain years (2013, 2014, and 2015). **Mahoba**: Displays a consistently small cultivated area compared to others, with a relatively smaller percentage decline. Overall, for the region**,** the year 2001 showed the largest cultivated area i.e.141,561 hectares and 2016 portrayed the smallest cultivated area i.e. 55,940 hectares (Fig. 14.2).

Analysis

The data suggests that sorghum is losing its importance as a cultivated crop in the Bundelkhand region, potentially due to:

- Changes in cropping patterns.
- Declining profitability or preference for other crops.
- Climatic challenges such as droughts or water scarcity.
- Reduction in demand for sorghum as a staple or fodder (Table 14.3).

The above data table provides information about **sorghum yield per hectare** across various districts in the Bundelkhand region of Uttar Pradesh from 2001 to 2022. We can analyse that the regional average yield has increased steadily over the years, achieving a **Compound Annual Growth Rate (CAGR) of 2.53%**. The **mean yield** across the entire period is **0.98 tonnes/hectare**, with a **standard deviation of 0.32**, indicating moderate variation. 2021 recorded the highest average yield of **1.65 tonnes/hectare**, showcasing strong performance across all districts and 2008 was the lowest-performing year, with an average yield of just **0.35 tonnes/hectare**, likely due to adverse climatic conditions. If we see district-wise **Jalaun** consistently ranks among the highest-yielding districts, especially in later years. **Jhansi** also shows strong improvements, peaking in 2022 with 2.28 tonnes/hectare. **Hamirpur** shows consistently lower yields, with only minor peaks in select years. **Mahoba**, though improving over time, has a low yield in several years, especially in 2016 (0.15 tonnes/hectare). **Notable Changes** in the above table are that there is a shift in yield

Table 14.2 Area in hectare for Sorghum cultivation in Bundelkhand Region of UP

Year	Jhansi	Jalaun	Chitrakoot	Hamirpur	Lalitpur	Mahoba	Total
2001	8082	14,643	28,685	36,657	5914	7916	141,561
2002	7524	14,573	25,996	35,318	4955	6854	128,585
2003	3497	10,123	15,877	31,614	3827	5854	102,224
2004	6430	12,650	21,883	34,113	7541	8662	125,410
2005	4373	8985	17,673	24,681	5342	5769	93,962
2006	4106	6225	18,960	20,933	2383	5667	82,817
2007	9200	11,791	16,072	28,590	4176	5207	101,713
2008	4564	10,657	17,196	20,471	1090	4085	80,856
2009	10,022	9105	14,864	15,529	4897	5907	82,315
2010	6590	9248	16,451	19,635	2543	6150	84,467
2011	3213	8954	15,584	22,648	931	4716	84,603
2012	2580	5243	16,819	19,213	566	4051	73,733
2013	1559	7244	18,046	18,074	75	3676	73,587
2014	1396	6656	12,794	15,537	86	3571	63,643
2015	1321	6810	12,446	15,783	14	1704	61,513
2016	294	6394	10,858	14,150	87	1158	55,940
2017	3263	7545	18,971	16,044	543	3918	74,086
2018	1934	6861	16,806	16,109	291	1936	67,652
2019	799	4048	16,093	8084	227	3130	54,900
2020	799	5471	18,212	13,049	209	2960	63,104
2021	1127	5523	17,144	16,620	581	2960	67,576
2022	1895	7700	19,716	14,732	247	3674	71,858
Total	**84,568**	**186,449**	**387,146**	**457,584**	**46,525**	**99,525**	**1,836,105**
Mean							**20799.27**
Std							**7665.03**
LOGEST							**0.97**
CAGR%							**−3.42296**

Source: 1. Directorate of Economics and Statistics, Dept. of Agriculture and Farmers Welfare, Ministry of Agriculture and Farmers Welfare, Govt. of India
2. CAGR based on Author's Calculation

from **2001 (0.82 tonnes/hectare)** to **2022 (1.58 tonnes/hectare),** which indicates a nearly **93% increase** in productivity over two decades. The largest improvement in yield occurred after 2013, potentially reflecting agricultural interventions or technology adoption (Fig. 14.3).

Analysis
The increasing productivity of sorghum in Bundelkhand is encouraging, especially given the challenges of water scarcity and marginal lands in the region. Contributing factors could include:

- **Government Support**: Implementation of schemes for dryland agriculture.
- **Improved Practices**: Adoption of drought-resistant varieties and modern farming techniques.
- **Farmer Adaptation**: Increased focus on yield optimization to counter shrinking cultivated areas.

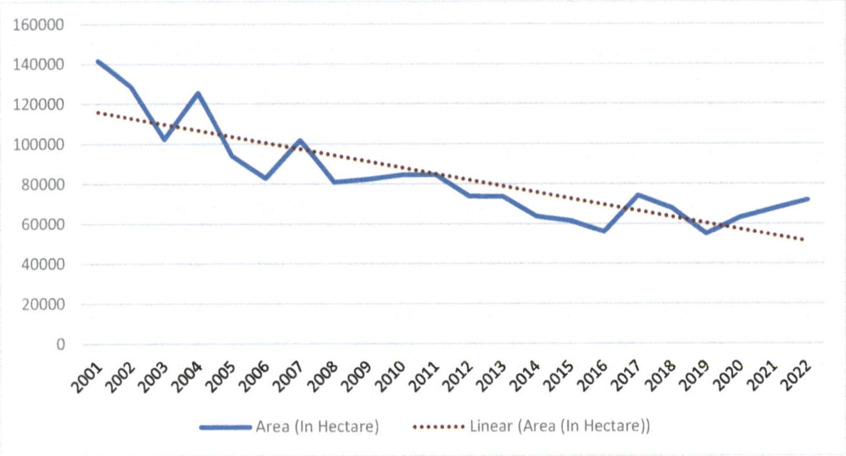

Fig. 14.2 Trend of area in hectare for Sorghum cultivation in Bundelkhand Region of UP

14.5 Results

The analysis of sorghum production in the Bundelkhand region of Uttar Pradesh from 2001 to 2022 reveals several key trends. The total production of sorghum has seen a slight decline, with a Compound Annual Growth Rate (CAGR) of −0.96%. In 2001, the total production was 118,809 tonnes, which decreased to 107,455 tonnes in 2022. District-wise, Banda showed the highest total contribution and a positive CAGR of 2.28%, while Lalitpur and Mahoba experienced the sharpest declines.

The area under sorghum cultivation has also decreased significantly, with a CAGR of −3.42%. In 2001, the total area was 141,561 hectares, dropping to 71,858 hectares in 2022, nearly a 50% reduction. The most significant decline occurred between 2001 and 2013, with a slower decline and slight stabilizations in some years post-2013.

Despite the reduction in cultivation area, the productivity of sorghum has increased, achieving a CAGR of 2.53%. The mean yield across the period is 0.98 tonnes/hectare, with notable improvements in later years. The highest average yield was recorded in 2021 at 1.65 tonnes/hectare, while 2008 was the lowest-performing year with an average yield of 0.35 tonnes/hectare. These trends indicate that while the area under cultivation has decreased, improvements in farming practices and technology have led to increased productivity.

Table 14.3 Tonnes per hectare for Sorghum Production in Bundelkhand Region of U.P

Year	Jhansi	Jalaun	Banda	Chitrakoot	Hamirpur	Lalitpur	Mahoba	Regions average yield
2001	0.86	0.98	0.85	0.84	0.81	0.64	0.79	0.82
2002	0.57	1.45	0.73	0.65	1.07	0.94	0.71	0.87
2003	0.2	0.8	0.65	0.64	0.86	0.63	0.7	0.64
2004	0.9	1.12	0.98	0.89	0.98	1.32	0.66	0.98
2005	0.9	1.35	0.9	1.1	0.81	0.89	0.87	0.97
2006	0.48	1.41	0.96	1.09	0.54	0.52	0.65	0.81
2007	0.97	1.68	0.74	0.63	0.55	1.57	0.87	1
2008	0.28	0.25	0.56	0.44	0.17	0.47	0.31	0.35
2009	0.87	1.84	0.49	0.77	0.41	1.16	0.8	0.91
2010	0.3	1.14	0.7	0.87	0.43	1.44	0.41	0.76
2011	1.28	1.11	0.98	0.83	0.73	1.23	0.89	1.01
2012	1.11	1.09	1.13	1.13	0.98	1.04	0.77	1.04
2013	1.93	1.9	1.43	1.24	1.32	1.11	0.97	1.41
2014	0.6	1.26	1.15	0.92	1.07	1.29	0.81	1.01
2015	0.87	0.86	1.11	0.91	0.82	0.86	0.55	0.85
2016	0.61	0.65	0.35	0.15	0.06	0.61	0.15	0.37
2017	0.95	0.95	0.92	1.13	0.67	0.95	0.96	0.93
2018	0.85	1.88	1.61	0.79	0.88	1.24	0.8	1.15
2019	1.71	1.41	1.38	0.85	0.85	1.7	1.02	1.27
2020	0.98	1.41	1.85	1.54	1.13	0.99	0.71	1.23
2021	1.23	1.64	2.38	1.85	1.02	1.59	1.82	1.65
2022	2.28	2.27	1.74	1.46	0.98	1.58	0.78	1.58
Total								**21.61**
Mean								**0.98**
Std								**0.32**
LOGEST								**1.03**
CAGR%								**2.53**

Source: 1. Directorate of Economics and Statistics, Dept. of Agriculture and Farmers Welfare, Ministry of Agriculture and Farmers Welfare, Govt. of India
2. CAGR based on Author's Calculation

14.6 Conclusion

The analysis of sorghum production in the Bundelkhand region of Uttar Pradesh reveals a complex picture. While the total production of sorghum has seen a slight decline, the productivity has increased significantly. The area under sorghum cultivation has decreased, but advancements in farming practices and technology have positively impacted productivity.

In light of the International Year of Millets 2023, sorghum production holds significant potential for the Bundelkhand region. Sorghum's resilience to drought and low input requirements makes it an ideal crop for the semi-arid conditions of Bundelkhand. Promoting sorghum cultivation can enhance food security, support sustainable agriculture, and contribute to the region's economic stability. The region's contribution to agriculture can be further justified by its ability to adapt to climatic challenges and optimize yield through improved practices.

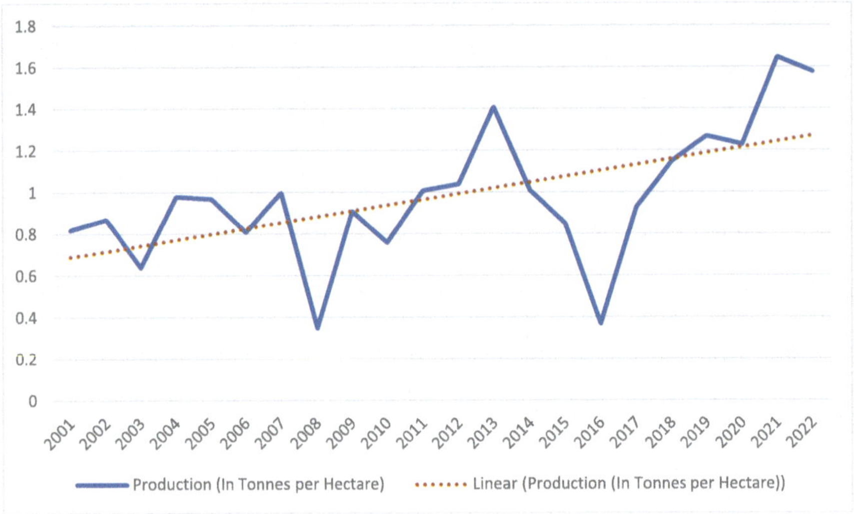

Fig. 14.3 Tonnes per hectare for Sorghum Production in Bundelkhand Region of U.P

By focusing on sorghum production, Bundelkhand can leverage its agricultural strengths, ensuring a stable food supply and contributing to the global recognition of millets as vital crops for sustainable development. Additionally, increasing sorghum production can help reduce food inflation by providing a reliable and affordable food source, thereby ensuring food security for India. This approach not only supports local farmers but also strengthens the overall agricultural economy.

References

Dalton, T. J., & Hodjo, M. (2021). Trends in Global Production, Consumption and Utilisation of Sorghum. In V. A. Tonapi, H. S. Talwar, A. K. Are, B. V. Bhat, C. R. Reddy, & T. J. Dalton (Eds.), *Sorghum in the 21st century: Food – Fodder – Feed – Fuel for a Rapidly Changing World* (pp. 3–15). Springer.

Karthick, R., Arulmani, K., & Ravi, K. (2023). Decline in the cultivation of minor millets in Tamil Nadu: A time series analysis of cultivation area, production and productivity (1952–53 to 2021–22). *Asian Journal of Agricultural Extension, Economics & Sociology, 41*(10), 432–442.

Pandey, N. L., Rath, R., Datta, M., Choudhury, B. R., & Kandpal, R. (2023, July). MAINSTREAMING MILLETS IN G20 COUNTRIES THROUGH INTEGRATED BEHAVIOURAL AND POLICY SHIFTS. *Task Force 6: Accelerating SDGs: Exploring NewPathways to the 2030 Agenda.*

Prakash, P., Mishra, D. K., & Mishra, D. (2024). ANALYSIS OF CLIMATIC VARIABLES AND ITS IMPLICATION ON AGRICULTURE IN THE BUNDELKHAND REGION OF UTTAR PRADESH, INDIA. *Journal of the Asiatic Society of Mumbai, 98*(10), 45–61.

RAI, S. K., BEHARI, P., SATYAPRIYA, RAI, A. K., & AGRAWAL, R. K. (2012). Long term trends in rainfall and its probability for crop planning in two districts of Bundelkhand region. *Journal of Agrometeorology, 14*(1), 74–78.

Sharma, D. H., Rathi, D., & Niranjan, D. K. (2020). *Strategies to bridge yield gap of major crops in Bundelkhand Region of India* (Jawaharlal Nehru Krishi Vishwa Vidyalaya, Jabalpur (M.P.) 482004, AGRO- ECONOMIC RESEARCH CENTRE). Ministry of Agriculture and Farmers Welfare, Government of India.

Singh, R. S., & Mishra, S. R. (2019, June). Trend and pattern of urbanization in Uttar Pradesh with special reference to Bundelkhand Region. *Geographical Review of India, 81*(2), 170–184.

Empowering Millet Entrepreneurs: A Focus on Sustainable Development Goals (SDG)

15

Uzma Fareen, Sushila Rajan,
and Mithileysh Sathiyanarayanan

Abstract

The United Nations General Assembly (UNGA) declared 2023 as the International Year of Millets to create awareness about the nutritional facts of millets, encourage entrepreneurs and investors to invest in research and development to improve the quality and sustainable production of millets. These crops are rich in nutrients and are wholesome foods as they provide a different and ecological source of nutrients. As part of the Sustainable Development Goals (SDGs) proposed by the United Nations, for addressing global issues and promoting sustainable development, it is essential to support millet cultivation. It not only helps promote economic growth but also eradicates hunger, promotes good health and well-being, provides clean water and sanitation, and contributes to industry, innovation, and infrastructure, as well as viable cities, communities, and climate action. This chapter identifies and highlights various challenges and opportunities to empower millet entrepreneurs.

Keywords

Millets · Entrepreneurs · Farmers · Empowerment · Sustainable development goals

U. Fareen · S. Rajan
Community Institute of Commerce & Management Studies, Bangalore, Karnataka, India

M. Sathiyanarayanan (✉)
Research & Development, MIT Square, London, Bengaluru, Karnataka, India

C. B. Singh et al. (eds.), *Global Millets Production for a Sustainable Future*,
https://doi.org/10.1007/978-3-031-88157-2_15

15.1 Introduction

The resurgence of millets as a nutritional powerhouse and a sustainable crop has sparked a wave of entrepreneurial endeavours worldwide. These small-seeded grains, once considered a staple but overlooked for years, are now gaining recognition for their immense potential in addressing various Sustainable Development Goals (SDGs).

Millets are the small seeded grains that are grown widely in the semi-arid areas of Asia and Africa. They belong to the family of Poaceae (i.e., grass family). Millet crops are highly favoured as they adapt well and grow even in dry and sultry conditions. The most commonly cultivated millets are Sorghum (Jowar), Pearl Millet (Bajra), Finger Millet (Ragi/Mandua) & Minor Millets namely: Foxtail Millet (Kangani/Kakun), Proso Millet (Cheena), Kodo Millet (Kodo), Barnyard Millet (Sawa/Sanwa/ Jhangora), Little Millet (Kutki) and two Pseudo -millets viz. Buckwheat (Kuttu) and Amaranthus (Chaulai).

Millets are of immense historical value as they are ancient grains that are cultivated for many generations. Traditional kitchens, customs, and values all reflect their cultural significance. Their capacity to adapt to a variety of climates helped them become widely distributed, supporting communities and influencing gastronomic traditions that demonstrated the intricate relationship between agriculture, culture, and history.

Millets play a major part in Indian cuisine. Millets are a very good source of energy as they release glucose, dietary fibre, a slow-digesting starch, and resistant starch (NAAS, 2013; Pradhan et al., 2009; Chetan & Malleshi, 2007; Hadimani & Malleshi, 1993). Millets contain fatty acids and are a great source of amino acids like cysteine and methionine. Millets are abundant in Vitamin E and Vitamin B and also minerals like calcium, phosphorus, magnesium, potassium, and iron. The nutrients in millets helps reduce the incidence of cancer, obesity, diabetes, cardiovascular diseases, gastric problems, migraines, and asthma.

The Government of India suggested to the United Nations that the year 2023 be titled the International Year of Millets (IYM-2023) in order to provide people with wholesome food and to generate demand both nationally and internationally. This proposal made by India was supported by 72 other countries.

On March 5, 2021, the United Nations General Assembly (UNGA) at its 75th session in declared 2023 as the International Year of Millets. Its main objectives were to increase awareness about the nutritional value of millets and encourage investors to invest in research and development to improve sustainable production and also the quality of millets. This chapter identifies and highlights various challenges and opportunities to empower millet entrepreneurs.

15.1.1 Why to Empower Millet Entrepreneurs?

The percentage of millet consumption is rising globally as people are educated about its nutrient contents and its adaptability to climatic conditions. As people nowadays are increasingly becoming health conscious they are trying to include foods like millet in their diets for their gluten-free nature.

To promote bio-friendly agricultural methods, preserve biodiversity, and eradicate global food insecurity, it is significant to empower millet entrepreneurs. Millet crops are considered an environmentally friendly choice as they require minimum water and adapt to a variety of climates. Empowering millet entrepreneurs helps in generating job opportunities, improving the lives of farmer and also in strengthening local economies. Furthermore, millets are nutrient-rich, providing numerous health benefits, and their consumption can solve nutritional deficiencies. Empowering millet entrepreneurs can help achieve viable development goals, improve health and lead to the economic growth of the country and also improve economic flexibility in communities around the world (ICAR, 2010; Nag et al., 2011; Ahmed & JulionCarign, 2012; Balasubramanium, 2014).

Micro loans or grants or advances can help millet entrepreneurs capitalize on infrastructure, equipment, and seed requirement for millet farming. Training programmes on sustainable farming methods, water management, and modern practices can be organized in rural areas. When there is access to larger marketplaces, businesses can reach a wider range of consumers, increase the demand for their products, increase prices, and promote economic growth.

Millet entrepreneurs can be motivated through cooperative farming, creating awareness about nutritional benefits, introducing various government schemes and policies, creating a sustainable market, and encouraging entrepreneurs to invest in millet cultivation.

15.1.2 How Does Promoting Millets Support Sustainable Development Goals (SDGs)?

As part of the Sustainable Development Goals (SDGs), to address global issues and promote ecological development, it is essential to support millet cultivation. It not only helps in promoting economic growth but also eradicates hunger, promotes good health and well-being, industry, innovation, and infrastructure, as well as eco-friendly cities and communities. Millets are rich in nutrients and are wholesome foods as they provide different and sustainable source of nutrients.

When millet entrepreneurs in rural areas are supported, there will be employment generation which leads to an increase in economic development (shown in Fig. 15.1). To promote millets, it is necessary to adopt the latest farming techniques and technologies and enhance the required infrastructure for its distribution. Millet cultivation helps achieve bio-friendly goals like poverty eradication, zero hunger, good health, well-being, responsible consumption and production, and climate action. Collaboration between governments, Non-Government Organizations (NGOs), enterprises, and communities is crucial to millet promotion which allows information exchange, resource sharing, and coordinated efforts to meet sustainability goals. In general, millets play a significant role in supporting sustainable development and tackling global climatic issues (Geetamma & Bulla, 2013; Global

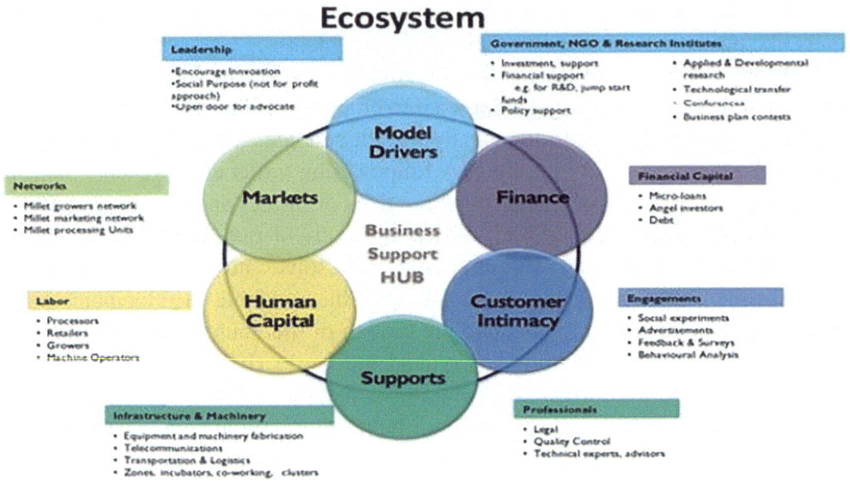

Fig. 15.1 The proposed ecosystem for the Indian small millets' agribusiness chain (Adekunle et al., 2018)

Entrepreneurship Monitor, 2007; Mamatha Rani & Chavan, 2019; Mathanghi & Sudha, 2012).

15.1.3 What Are the Challenges and Opportunities for Millet Entrepreneurs?

Millet Entrepreneurs have to face numerous obstacles, such as limited market access, less awareness about the nutritional benefit of millets, technological challenges, susceptibility to climate change, and lack of government support. The unavailability of modern agricultural technology and tools might affect yield and efficiency and also variable weather patterns can make steady farming challenging. The growth of millet entrepreneurship can also be hindered by unclear regulations or insufficient policy funding which can have an impact on market restrictions, land use policies, and subsidies. All millet entrepreneurs have to overcome these obstacles to meet success in business (ICAR, 2010; Nag et al., 2011; Ahmed & JulionCarign, 2012; Balasubramanium, 2014).

The increasing demand for sustainable agriculture, the beneficial trend of healthy eating, the prospects for value addition by developing millet products like flour, snacks, or products tailored to dietary requirements, various government schemes and initiatives supporting sustainable agriculture, and the potential for exotic foods in global markets are some of the opportunities that millet entrepreneurs can exploit. An atmosphere that supports sustainable agriculture through policies, subsidies, and initiatives from the government can also be favourable for millet businesses. Global markets provide millet entrepreneurs a chance to take advantage of the growing craze for unique and rare cuisines. Businesses can reach a new high if entrepreneurs

Table 15.1 Challenges and opportunities for millet entrepreneurs

Challenges for millet entrepreneurs	Opportunities for millet entrepreneurs
1. Market access, awareness and penetration	1. Rising health and sustainability awareness
2. Agricultural constraints	2. Innovation and product diversification
3. Policy and regulatory issues	3. Technological advancements
4. Perception and consumer preferences	4. Advocacy and education
5. Infrastructural constraints	5. Supportive policies and partnerships
6. Technological challenges	6. Consumer education and marketing
7. Export barriers & local challenges	7. Collaborative initiatives
8. Quality control and eco-friendly packaging	8. Sustainable Agriculture and climate resilience
9. Human capital challenges	9. Digital expansion and global reach
10. Accessibility & transportation challenges	10. Investing in millet prosperity & fresh perspectives

capitalize on research and development for various millet growing methods, product creation, and marketing techniques. But to overcome these obstacles it requires a planned strategy, teamwork, flexibility, and the creation of associations and networks within the agricultural network. The challenges and opportunities for millet entrepreneurs we identified are given in the table below (Table 15.1) and explained in detail in the below sections.

15.1.4 How Self-Help Groups (SHG) Can Play a Role in Empowering Millet Entrepreneurs?

Self-Help Groups (SHGs) can be a great support by providing access to capital, skill development, community support, market networks, risk moderation, advocacy and representation, and women's empowerment. SHGs give millet entrepreneurs a way to combine their resources and get access to financial services including microloans and group savings accounts. They also provide a platform where entrepreneurs can share their knowledge, enhance their skills, counsel, get emotional support, and share some tactics or strategies for strengthening their businesses. Through collective bargaining, millet entrepreneurs can solve issues such as price fluctuations and market accessibility while negotiating higher pricing for their produce. SHGs also help business owners to create a network that will enable entrepreneurs to reach a wider audience.

SHGs serve as a means of risk sharing, offering assistance in the event of crop failures or unforeseen complications. SHGs can operate as a consolidated representation for millet entrepreneurs, promoting their concerns and influencing judicial choices to foster an environment that supports millet enterprise and agriculture. SHGs can empower women through millet entrepreneurship in the agricultural industry. By working together and supporting one another, SHGs can assist millet entrepreneurs in the long-term development of millet-based businesses (Mohamed et al., 2012; Okpala, 2012; Patankar & Metha, 2014; Pradhan & Nath, 2012).

15.2 Millet Entrepreneurship from Sustainable Development Goals Point of View

Millet entrepreneurship is not just about cultivating and selling these grains; it's a movement that encompasses diverse aspects of sustainable development, aligned with multiple SDGs set by the United Nations.

15.2.1 Addressing Hunger and Food Security (SDG 2)

Millet, with its high nutritional value, contributes significantly to food security. It's resilient in harsh environments, requiring minimal water and fertilizer compared to other grains. Empowering millet entrepreneurs means promoting a crop that can thrive in diverse climates, providing sustenance where conventional crops struggle.

15.2.2 Promoting Health and Well-being (SDG 3)

Rich in nutrients like iron, calcium, and fibre, millets combat malnutrition and diet-related diseases. By encouraging the consumption and production of millets, entrepreneurs actively contribute to public health, offering an alternative to processed foods and fostering healthier eating habits.

15.2.3 Sustainable Agriculture (SDG 12)

Millet cultivation often follows traditional, sustainable agricultural practices. Its low water requirement and ability to grow in poor soil reduce the carbon footprint. Supporting millet entrepreneurs is endorsing environmentally friendly farming methods, and promoting ecological friendly agriculture.

15.2.4 Gender Equality and Economic Growth (SDG 5 and 8)

In many regions, women are involved in various stages of millet production. Empowering these women entrepreneurs not only contributes to economic growth but also fosters gender equality by providing opportunities for income generation and leadership roles in agricultural communities.

15.2.5 Climate Action (SDG 13)

Millet's pliability to drought and its ability to withstand high temperatures make it a climate-resilient crop. By promoting its cultivation, entrepreneurs actively mitigate climate change reduce the vulnerability of agricultural systems to hostile weather conditions.

15.2.6 Preserving Biodiversity (SDG 15)

The cultivation of millets helps to preserve biodiversity by promoting indigenous crops. By supporting millet entrepreneurs, we encourage the preservation of diverse seed varieties and traditional farming practices, contributing to the conservation of biodiversity.

15.3 Challenges for Millet Entrepreneurs

Millet entrepreneurs are at the forefront of a movement promoting sustainable agriculture, nutrition, and economic growth. However, they face several challenges in their endeavours to popularize and scale the production and consumption of these nutritious grains. Some of them are mentioned below:

15.3.1 Market Access, Awareness and Penetration

- The lack of widespread knowledge about millets among consumers poses a significant hurdle. Many are unfamiliar with their nutritional benefits and gastronomic versatility.
- Limited market access and distribution channels restrict the reach of millet-based products, making it harder for entrepreneurs to establish a foothold in the industry.

15.3.2 Agricultural Constraints

- Millet cultivation often relies on traditional methods, limiting scalability and efficiency.
- Climate variability and inadequate infrastructure for irrigation and storage can affect yields and quality.

15.3.3 Policy and Regulatory Issues

- Inconsistent policies and regulations related to pricing, subsidies, and certification standards can hinder the growth of millet entrepreneurship.
- Insufficient government support and incentives may discourage farmers from transitioning to millet cultivation.

15.3.4 Perception and Consumer Preferences

- Prevailing consumer preferences for more familiar grains and processed foods create a challenge in popularizing millets.
- Perceived as 'poor man's food' in some cultures, this stigma requires creating awareness and innovative marketing strategies.

15.3.5 Infrastructural Constraints

- Processing Facilities: Insufficient infrastructure for processing and value addition hampers scalability and quality control.
- Storage Facilities: Inadequate storage facilities lead to post-harvest losses, affecting the food supply chain.

15.3.6 Technological Challenges

- The effective manufacture of millet products may be disadvantaged by lack of access to cutting-edge processing technologies.
- The effectiveness of millet cultivation may be affected by limited access to contemporary agricultural technologies and machinery.

15.3.7 Export Barriers and Local Challenges

- Trade laws and export limitations pertaining to millet products can present challenges for entrepreneurs seeking to reach global markets.
- Local challenges like political instability can provide difficulties for entrepreneurs and have an impact on the general business climate that relies on millet.

15.3.8 Quality Control and Eco-Friendly Packaging

- It can be difficult to maintain the same quality between batches which might reduce the credibility and confidence people have in millet-based products.
- It might be difficult to create packaging that protects the quality and freshness of millet goods using eco-friendly materials.

15.3.9 Human Capital Challenges

- As millet is a seasonal crop, it can be difficult to find reliable, competent workers at busy times.
- In certain areas, it may be difficult to find experienced labourers who are extremely knowledgeable about growing millets.

15.3.10 Accessibility and Transportation Challenges

- Ensuring the availability of millet-based products in remote areas can be logistically challenging, affecting market reach.
- As there are inadequate transport facilities to remote areas, the delivery of these products may get delayed and increase the cost of distribution.

15.4 Opportunities for Millet Entrepreneurs

Millet entrepreneurs are at the forefront of a burgeoning movement aimed at capitalising various opportunities to make their businesses powerful and successful. Here are some key opportunities we have identified.

15.4.1 Rising Health and Sustainability Awareness

- Increasing awareness of health and sustainability benefits fuels the demand for nutritious and eco-friendly foods, creating a ripe market for millets.
- The growing interest in gluten-free and organic products positions millets as an attractive alternative.

15.4.2 Innovation and Product Diversification

- Entrepreneurs can capitalize on innovation in processing techniques to create a diverse range of millet-based products like flour, snacks, beverages, and ready-to-eat meals.
- Collaborations with chefs, nutritionists, and food scientists can help develop appealing and convenient millet-based recipes.

15.4.3 Technological Advancements

- Adoption of modern agricultural technologies, such as precision farming and mechanization, can enhance millet cultivation efficiency and yield.
- Research and development in crop breeding for improved varieties resistant to pests and climate fluctuations can bolster production.

15.4.4 Advocacy and Education

- Educational campaigns and advocacy efforts can raise awareness about millets' nutritional value and their role in sustainable agriculture, impacting consumer behaviour positively.

- Collaborations with influencers, nutritionists, and health experts can help reform perceptions around millets.

15.4.5 Supportive Policies and Partnerships

- Encouraging policy frameworks, including subsidies, market support, and incentives, can incentivize farmers and entrepreneurs to invest in millet production and processing.
- Partnerships between governments, NGOs, and private sectors can provide technical assistance, access to finance, and market linkages for millet entrepreneurs.

15.4.6 Consumer Education and Marketing

- Health and Sustainability: Rising health consciousness and environmental concerns present an opportunity to position millets as nutritious and eco-friendly alternatives.
- Innovative Marketing: Utilizing digital platforms and partnerships for targeted marketing campaigns can increase awareness and demand.

15.4.7 Collaborative Initiatives

- Public-Private Partnerships: Collaboration between governments, NGOs, and private sectors can create supportive ecosystems with training, infrastructure, and policy advocacy.
- Value Chain Integration: Strengthening links across the value chain through cooperatives or associations can streamline production, processing, and distribution.

15.4.8 Sustainable Agriculture and Climate Resilience

- Environmental Concerns: Growing awareness of climate change fosters interest in climate-resilient crops like millets, promoting sustainable agricultural practices.
- Certifications and Labels: Certifications for organic and sustainable farming can enhance market access and premium pricing.

15.4.9 Digital Expansion and Global Reach

- In the fast-developing technological world, millet entrepreneurs can utilize various digital or online channels to expand their customer base.

- Millet Entrepreneurs can explore foreign markets, especially those where people are looking for a variety of healthy food options.

15.4.10 Investing in Millet Prosperity and Fresh Perspectives

- Millet Entrepreneurs can invest in developing weather forecasting software or technology which will help reduce risks associated with unpredicted weather patterns and they can also invest in cutting-edge irrigation technologies which will help overcome water scarcity issues.
- They can also initiate to invest in mobile apps for farmer education which guides the farmers on the best methods to get higher quantity and quality of millet crops.
- Establishing farm-to-table models can attract consumers looking for fresh, locally sourced products and improve supply chain transparency.
- Exploring partnerships with sectors such as health, beauty, etc., can result in the discovery and creation of unique products using millets.
- Showcasing millet products in food festivals and events can provide exposure and networking opportunities within the food industry.

15.5 Conclusion

Millets –small seeds packed with multivitamins and minerals are considered a superfood with health benefits with its demand increasing globally. Embracing millet as a key player in the entrepreneurial sphere to nourish the community, and empower individuals can create a strong and extensive future for agriculture and the food system.

Empowering millet entrepreneurs is not just about promoting a single crop; it's about endorsing a sustainable lifestyle that aligns with multiple SDGs. It's a step towards a more resilient, inclusive, and healthier future for communities worldwide. Governments, NGOs, and businesses must collaborate to provide the necessary support, including financial assistance, technical knowledge, and market access to foster a successful network for millet entrepreneurs. This collective effort can pave the way for a more sustainable and equitable world while harnessing the potential of a humble yet powerful grain—the millet.

While challenges persist, the landscape for millet entrepreneurs is evolving favourably. Addressing these challenges through innovation, education, policy support, and strategic collaborations can unlock the immense potential of millets, not just as a nutritious crop but also as a catalyst for sustainable development and economic empowerment in communities around the world.

Awareness programs to spread information about the nutritional benefits of millets to the consumers should be increased: farmers, rural women, and entrepreneurs should be encouraged to carry out millet-based businesses and training programs should be introduced which will be helpful in increasing their knowledge about cultivating millets and running their business successfully. The day that millet entrepreneurs run their business profitably the economy of the country will soar. In this scenario, jobs are generated, rural women are empowered and the health of a nation improves through the enhanced nutritional consumption of people (Sathiyanarayanan et al., 2019; Almutairi & Sathiyanarayanan, 2015).

References

Adekunle, A., Lyew, D., Orsat, V., & Raghavan, V. (2018). Helping agribusinesses—Small millets value chain—To grow in India. *Agriculture, 8*(3), 44.

Ahmed Zafer, U., & JulionCarign, C. (2012). International entrepreneurship in Lebanon. *Global Business Review, Sage Publication, 13*(1), 25–38.

Almutairi, F., & Sathiyanarayanan, M. (2015). Causes of failure of implementation management information systems (MIS) in small and medium enterprises (SMEs) in the Middle East. *International Journal of Innovative Research and Creative Technology, 1*(1), 32–35.

Balasubramanium, S. (2013). Processing of millets—Lead Paper. Conference: National Seminar on Recent Advances in processing, utilization and nutritional impact of small millets, At I, 1–14, Volume: I, 1–14. Fatoki, O. (2014). The financing options for new small and medium enterprises in South Africa. *Mediterranean Journal of Social Sciences, 5*(20), 748.

Chetan, S., & Malleshi, N. G. (2007). Finger millet polyphenols: Optimization of extraction and effect of pH on their stability. *Food Chemistry, 105*, 862–870.

Geetamma K. and Bulla, B. (2013). Golden research thoughts.

Global Entrepreneurship Monitor. (2007). *2006 report on women and entrepreneurship*. Babson.

Hadimani, N. A., & Malleshi, N. G. (1993). Studies on milling, physicochemical properties, nutrient composition and dietary fiber content of millets. *Journal of Food Science and Technology, 30*, 17–20.

ICAR. (2010). ICAR news—A science and technology newsletter. *Indian Council of Agricultural Research, New Delhi, India, 16*(3), 16.

Mamatha Rani, R., & Chavan, U. D. (2019). Kotecha PM and Lande SB preparation and storage study of millet puttu. *International Journal of Chemical Studies (IJCS), 7*(3), 4453–4457.

Mathanghi, S. K., & Sudha, K. (2012). Functional and phytochemical properties of finger millet (Eleusine coracana L.) for health. *International Journal of Pharmaceutical, Chemical and Biological Sciences, 2*(4), 431–438.

Mohamed, T. K., Issoufou, A., & Zhou, H. (2012). Antioxidant activity of fractionated foxtail millet protein hydrolysate. *International Food Research Journal, 19*(1), 207–213.

NAAS. (2013). *Role of millets in nutritional security of India* (Policy Paper No. 66). National Academy of Agricultural Sciences, 16 p.

Nag, S. K., Mishra, N. K., Patel, S., Patil, S. K., Rao, S.S., Verma, P.K., Sahu, R. And Netam, R.S. (2011). "Empowerment to village entrepreneurship through tamarind—Processing techniques". In: Engineering interventions in agriculture—The way to improve rural economy, Indira Gandhi Krishi Vishwavidyalaya, , pp. 608–610.

Okpala, K. E. (2012). Venture capital and the emergence and development of entrepreneurship: A focus on employment generation and poverty alleviation in Lagos state. *International Business and Management, 5*(2), 134–141.

Patankar, V., & Metha, N. K. (2014). A literature review on entrepreneurial development and its antecedents. *Conference: 3rd Annual Commerce Convention on Leveraging Business: Discovering New Horizons, At School of Economics*, New Delhi.

Pradhan, R., & Nath, P. (2012). Perception of entrepreneurial orientation and emotional intelligence: A study on India's future techno-managers. *Global Business Review, Sage Publications, 13*(1), 89–108.

Pradhan, A., Nag, S. K., & Patil, S. K. (2009). Dietary management of finger millet (Eleusine coracana L. Gaerth). *Current Science, 98*, 763–765.

Sathiyanarayanan, M., Junejo, A. K., & Fadahunsi, O. (2019). Visual auxiliary solutions to analyse social media data for improving marketing & business. In *2019 International Conference on contemporary Computing and Informatics (IC3I)* (pp. 169–174). IEEE.

Promoting Millet-Based Products for Enhanced Health and Immunity: Towards a Resilient World

16

R. M. Chitnis and Vijay Kulkarni

Abstract

Nutrient-rich grains are fast making a resurgence in Indian agriculture after decades of institutional neglect. These cereals were disregarded by development groups and farmers who preferred to grow rice, wheat, and other crops like oilseeds and pulses. When growing on poor soil, millets need less water, fertilizer, and pesticides to thrive. Because they can withstand higher temperatures, they are the best option for "climate-smart" cereals. China, India, and Nigeria are the world's top producers of millets, accounting for about 55% of global production. Millets are a good choice since they are high in antioxidants, vitamins, and minerals. Common cereals like millets may be a good source of nutrients that support immunity. This is particularly relevant in view of the ongoing Covid-19 virus outbreak. The purpose of this research is to ascertain the customers' knowledge level that motivates them to use goods based on millet or millets. There were 220 responders in the survey. According to the study's results, the majority of them planned to include millets in their regular diet since they were aware of their health advantages.

Keywords

Millet-based products · Health promotion · Immunity enhancement · Resilient world

R. M. Chitnis (✉)
MIT World Peace University, Pune, Maharashtra, India
e-mail: vc@mitwpu.edu.in

V. Kulkarni
Ajeenkya DY Patil University, Pune, Maharashtra, India

C. B. Singh et al. (eds.), *Global Millets Production for a Sustainable Future*,
https://doi.org/10.1007/978-3-031-88157-2_16

16.1 Introduction

Known as needy individuals' food, millets are little cultivated grasses that are filled in parched zones of Asia and Europe. They are more affordable than wheat and rice. With regards to proteins, minerals, and nutrients, they are three to multiple times more nutritious than staple grains. Only traditional consumers and those from lower socioeconomic classes can afford the nutritionally superior millets. However, the unpredictability of the weather and the world's population growth force nutritionists to create healthy, functional millet-based foods that not only combat poverty and hunger but also gain widespread public acceptance. Micronutrients found in millets, such as vitamins and β-carotene, are abundant and are currently consumed similarly to pharmaceutical pills. Millets are supplement thick with sweet and nutty flavour, are effectively edible and non-unfavourably susceptible grains accessible for utilization alongside their illness resistance to be developed under exceptionally antagonistic circumstances with less work. Because of their higher mineral content, they alkalize the body, which lowers the incidence of lifestyle diseases. Since millets are high in dietary fiber, safe starches, nutrients, fundamental amino acids, stockpiling proteins, and other bioactive substances, they are healthfully better than other significant oats. Little millets are excellent crops for climate and emergency plantings because of their high resilience to adjust to various ecological conditions. Millets' special nutritional and nutraceutical qualities contribute to the global provision of nutritional security to a greater number of underprivileged and economically vulnerable populations.

16.1.1 Climate Resilient Crops

Millets are sensitive to environmental pressures and may adapt to a broad variety of ecological situations. They need less irrigation and grow and produce better in low-nutrient settings with no need for chemical fertilizers. One of the main policy objectives is to provide food security for the country. Millets are rich in proteins, fibre, B-complex vitamins, lecithin, the sulphur-containing amino acid methionine, and very little vitamin E. Iron, magnesium, calcium, and potassium are abundant in them. Phytonutrients in the seeds also decrease cholesterol and diminish the risk of cancer. Due to over exploitation of agricultural areas with irrigation capabilities, the world is now facing nutritional and agrarian issues. Focusing on dry lands is essential to boost grain production for the world's rapidly growing population, yet producing enough grains on these sparsely fertilized plains is a significant difficulty. The following qualities make millet a very significant crop: it has a shorter growing season than other main grains, resilience to pests and illnesses, and a reputation for being drought-tolerant. When grown in poor settings, millets have more nutritional content than staple grains and are considered a climate change compliant crop. Nutrients, minerals, fundamental unsaturated fats, phytochemicals, and cell reinforcements remembered for these Nutri-grains might help with the destruction of a great many nourishing issues. Its cultivation can maintain the productivity of arid environments and guarantee future food security. Reviving millet farming in Telangana, Karnataka, and AP might be a step toward sustainable agricultural methods that support biodiversity. Because millets use about 2.5 times less water than rice, according to ICRISAT, millets are more sustainable as crops for a variety of reasons.

16.1.2 Superior Grains for Health

Climate change, water and food shortages, an ever-increasing global population, growing food costs, and socioeconomic issues are all posing threats to global agricultural and nutrition security in the twenty-first century. Since millets are supplement thick, high in cell reinforcements, basic framing, sans gluten, and versatile to different soil and climatic circumstances, they might be a brilliant food source to assist with combatting craving and destitution while additionally guaranteeing food security for those living in dry and semi-parched regions. It is feasible to utilize food handling methodology to build the healthy benefit of food, decline enemies of parts, and work on the assimilation and bioavailability of dietary supplements. Millets give a few advantages to food, sustenance, grub, fibre, health, business, and the climate in India. The edible stalks of most millets are preferred for cow feed and as nutritional stores for them. The basic tenet of traditional medicine was that prevention of illness is more important than treatment, with food being both medicine and medication considered nourishment. According to this method, using dietary supplements in moderation helps shield humans from terrible illnesses. Because of its medicinal potential, navadhanyam, or cereals and millets, are essential for providing for the unique requirements of patients recuperating from disease as well as expectant and nursing mothers. For an expansion in dietary fibre, nutrients, and minerals, most of grains and heartbeats might be splashed for the time being and eaten as fledglings. Because of their high nutritional content, millets may help prevent cancer, lower blood pressure, heart disease risk, lower the incidence of cardiovascular diseases, slow down the growth of tumours, increase gastrointestinal bulk, delay gastric emptying, and lower cholesterol. Millets are non-glutinous grains that may be used in place of wheat in diets by those with gluten allergies and celiac disease. Because millets don't produce acid in the digestive system, they are non-allergic and simple to digest. Millets help individuals fight lifestyle illnesses since they are free of gluten, high in antioxidants, and contain waxes that decrease cholesterol. They may provide millets-producing farmers a market and save the foreign currency needed to buy basic grains. The increased need for gluten-free goods in today's society due to the prevalence of celiac illness.

16.2 Research Objectives

- To ascertain the degree of consumer knowledge about millet.
- To examine the factors influencing consumers' use of millet.
- To look at how the pandemic has affected the consumption and buying of millet.

16.3 Literature Review

Samtiya, M., Dhaka, N., Dhewa, T., Puniya, A. K., and Aluko, R. E. (2023). This survey frames a few gainful dietary properties of millet seeds and their subordinates for human health, including their job as stomach health modulators, cell reinforcement, immunomodulatory or mitigating, antibacterial or antimicrobial, hypocholesterolaemia, hypoglycaemic, and hostile to cancer-causing potential. Despite the fact that there are numerous variations, this concentrate generally centres around pearl millet (Cenchrus americanus, ordinarily known as Pennisetum glaucum), which is one of the most famous millet crops cultivated in India. Other millet species are additionally examined. The regular fixings (proteins, peptides, polyphenols, polysaccharides, oil, isoflavones, and so on) found in millet seeds are analysed in this article for their potential health benefits. Human mediation taking care of preliminaries are important to approve various of the conceivable health benefits of millet seeds, regardless of whether a large number of them have been shown involving creature model's in vitro research. Finger millet and foxtail millet are suggested as reasonable choices for consideration in future dietary medicines for upgraded human health, based on the nourishing and health-advancing characteristics related with pearl millet (canvassed in this article).

Islam, N., Khan, H. H., Sabuz, A. A., Rana, M. R., Ahmed, T., Molla, M. M., & Shen, Q. (2023). Millet is an important source of animal feed, and its production has increased dramatically in recent years to meet the growing nutritional needs of the global population. It is an excellent source of vital nutrients, including protein, fat, carbohydrates, minerals, and vitamins. It also contains a few other bioactive compounds that have a variety of biological effects, such as regulating the composition of the gut microbiota, acting as an antioxidant, lowering cholesterol, lowering blood pressure, and having an impact on anthropometric measurements. Handling techniques like decortication, splashing, malting, processing, maturation, and others could affect these bioactive parts, supplements, and uses of cereal grains. Based on roughly 150 distributions distributed somewhere in the range of 2015 and 2022 from the Snare of Science, Google Researcher, Food and Farming Association of the Assembled Countries (FAO), Rearing Bid Overview (BBS), and Food Information Focal (USDA) data sets, this study looks at the dietary and utilitarian characteristics of millet-consolidated food sources and their impacts on health. Surveying the examination clarifies that adding millet and its parts to eat less carbs could assist with forestalling undernourishment and various different ailments. Moreover, this examination offers significant subtleties on the health advantages of millet, which might be utilized as a norm for business, shoppers, scholastics, and nutritionists.

Tripathi, G., Singh, B. V., Nath, D., Das, H., Bansal, S., Borah, A., Jitendra Kumar, P. H., & Singh, B. V. (2023). Millets are prized for their nutritional value, which includes vitamins, minerals, carbs, and dietary fibre sources, as well as possible health benefits. Recent years have seen a rise in interest in millets because of their exceptional nutritional makeup and possible health benefits. This review looks at the many varieties of millets, their cultural distribution, and their historical relevance. Because of their high nutritional content, which includes proteins, dietary fibre, vitamins, and minerals, millets are a great complement to a diet that is well-balanced. Moreover, it delves into the gluten-free qualities of millets, accommodating those with particular dietary needs. The potential of millet intake to prevent and manage chronic illnesses such as diabetes and cardiovascular ailments is the main topic of discussion about its health benefits. The potential contribution of bioactive chemicals, such polyphenols and antioxidants, to these health advantages is being studied. The study also covers millets' effects on intestinal health and their potential contributions to sustainability and global food security in the future. All things considered, the results highlight millet as a wholesome and viable alternative for improving human health.

Millets are a remarkable wellspring of solvent and insoluble dietary filaments, minerals, cell reinforcements, and safe starch. As indicated by Ali et al. (2003), it has around 92.5% dry matter, 2.1% debris, 2.8% rough fibre, 7.8% unrefined fat, 13.6% rough protein, and 63.2% starch. As indicated by Glew et al. (2008), dark finger millet has 8.47 g/g of dry weight protein and 8.71 mg/g of unsaturated fat. As indicated by Hegde and Chandra (2005), kodo millet and little millet contain the best degrees of dietary fibre among all oats, going from 37% to 38%. Additionally, the fat content of these grains is higher in polyunsaturated fatty acids (PUFA). Proso millet has a protein level of around 11.6%, according to Kali nova and Moudry (2006). It is high in important amino acids, namely leucine, isoleucine, and methionine.

As per research by Chaturvedi et al. (2011), the most well-known oat based practical food sources and nutraceuticals are produced using wheat rice, millets, grain, oat, buckwheat, maize, sorghum, flaxseed, psyllium, and earthy coloured rice products. Cereal supplements have been displayed to possibly bring down pulse, the gamble of malignant growth, diabetes, coronary illness, cancer occurrence, and assimilation of fat and cholesterol. They may also delay the emptying of the gastrointestinal tract and promote gastrointestinal health. Cereals and their processed derivatives may contribute to disease prevention and health promotion when consumed on a regular basis. The lessening sugar content, all out starch, natural investigation, dampness content, complete fat, all out protein, absolute fibre content, and all out debris were determined by Sandeep et al. (2014) based on their general and phytochemical assessment of the minor millet seed coat.

K. K. Anagha (2023). Pseudocereals like millet, which are mostly cultivated in Africa and Asia, are becoming more and more popular in North America as a result of rising consumption among diasporic South Asian and African populations. Leading the world in millet production, India grows important kinds such as Finger Millet, Pearl Millet, and Sorghum. Compared to other cereals, millets have a number of benefits, including as resilience to diseases and pests, quicker maturation times, and production in dry environments. They are particularly well-nourished, having more calcium, dietary fibre, and protein than other foods. In addition, millets are devoid of gluten, which means that those who have gluten sensitivity or celiac disease might benefit greatly from them. Millets are surprisingly high in phytochemicals and antioxidants, which provide a host of health advantages including lowering blood pressure, avoiding diabetes, cancer, and heart disease, and boosting immunity. The nutritional qualities of millets are further improved by conventional processing techniques, making them appropriate for industrial use and the production of a variety of food items with added value. The purpose of this study is to draw attention to the remarkable properties of millets and to illuminate their potential as a highly nutritious and sustainable food source, particularly in the North American market.

16.4 Research Methodology

The purpose of the study was to learn how respondents felt about millet and ready-to-eat goods made from millet. The COVID-19 epidemic increased people's awareness of their health and encouraged them to eat millet. A lot of hotels have opened to highlight their own millet, particularly those that provide "Millet Based Food."

The purpose of this research is to gauge consumer knowledge of millet and items made from it. Through social media, a structured questionnaire was sent to responders from throughout India. Almost 220 people took part and gave their thoughts. Using SPSS, the gathered data were examined.

16.5 Data Analysis

The respondents' demographic profile is the first section of the study. Due to the fact that it is a linked food item, women are more worried than men, making up 70% of the responses. Fifty percent of individuals surveyed are under thirty, and fifteen percent are between thirty and forty. Of the respondents, 14% are between the ages of 41 and 50, while the remaining respondents are older than 50.

The typical month to month pays of the 13% of respondents was not as much as Rs. 20,000. A fifth of them procure between Rs. 21,000 and Rs. 35,000 consistently. Of them, 40% make between Rs. 36,000 and Rs. 50,000, while the excess people procure more than Rs. 50,000. Data for buyers comes from various sources. The information source's Friedman-Test result is shown in the table below (Tables 16.1, 16.2 and 16.3).

The results showed that "social media" accounted for the bulk of the information that consumers got. It serves as the information centre. Second, the finest source of information, in the respondents' opinion, is word of mouth. When it comes to sources of knowledge and awareness regarding millet, TV, newspapers, magazines, seminars, and conferences rank third, fourth, fifth, and sixth (Figs. 16.1, 16.2 and 16.3).

With millet-based health drinks receiving the top score and millet-based snacks receiving the second highest ranking, people are becoming more conscious of their health. Noodles made from millet, Dosa/Idly mixture, infant food, and cereals all received the same scores of third, fourth, fifth, and sixth.

Table 16.1 Awareness-raising sources: millet and goods made from it

Sources	Mean	SD	Mean rank	Rank
Social media	3.59	1.430	5.71	1
Friends/ relatives	3.60	1.060	4.13	2
Online shopping portals	2.91	1.200	3.97	3
TV	2.23	1.510	2.97	4
Newspapers/ magazines	1.65	0.640	2.93	5
Seminars/conferences	1.11	0.954	2.91	6

Table 16.2 Knowledge of and inclination for millet and items made from millet

Sources	Mean	SD	Mean rank	Rank
Millet health drink/ mix/porridge	4.32	1.340	4.20	1
Millet snack/ energy/granola bars/cookies/flakes/ pan cake	3.29	1.250	3.75	2
Millet noodles/ pasta	3.33	1.700	3.25	3
Millet dosa/idly mixes	4.23	0.998	3.11	4
Millet baby food	3.11	1.280	3.01	5
Millet breakfast cereal/muesli	3.01	0.947	2.87	6

16.6 Conclusion

Every single person is accountable for the development of sustainable practices; they are the ones who need to start moving in the right direction for the environment. Millets are beneficial to both agriculture and the environment thanks to their consumption. Most of them are eating low quality food, which adds to an unhealthy environment; thus, people need to change the ways of behaving that they have about their admission of food. These days, there are a lot of hotels and restaurants that are centered on millet. On the internet, you can only find a very limited selection of millet-based food products. Online millet food products may be provided by restaurants and motels because to the fact that people have a natural tendency to seek for a wide selection of options. Consumption of millet has been shown to be associated with the development of an environment that is immune-boosting, healthy, and full of vitality among millet eaters.

16.6.1 Recommendations

- Invest in social media marketing strategies that are specifically targeted in order to capitalize on the popularity of these platforms in order to promote goods that are based on millet.
- Taking into consideration the desire for online shopping portals as a source of information, it is recommended that millet-based goods be expanded to include a wider presence on different online platforms.
- To increase knowledge about the nutritional advantages of millets, educational efforts should be conducted both online and offline. These campaigns should include the participation of nutritionists or health professionals.

Table 16.3 Millets consumed both before and after COVID 19

Consumption frequency	Before COVID 19%	After COVID 19%
Once in a day	10	60
Once in a week	30	30
Monthly once or occasionally	60	10

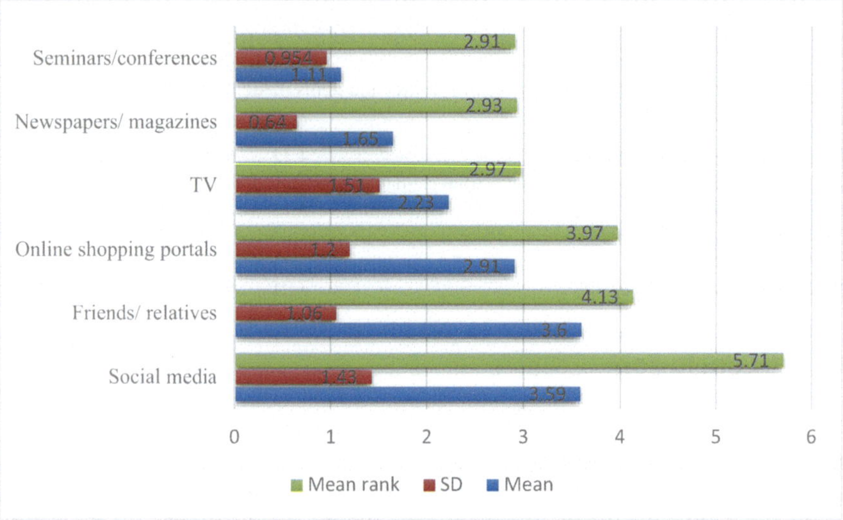

Fig. 16.1 Graphical represent of sources of awareness

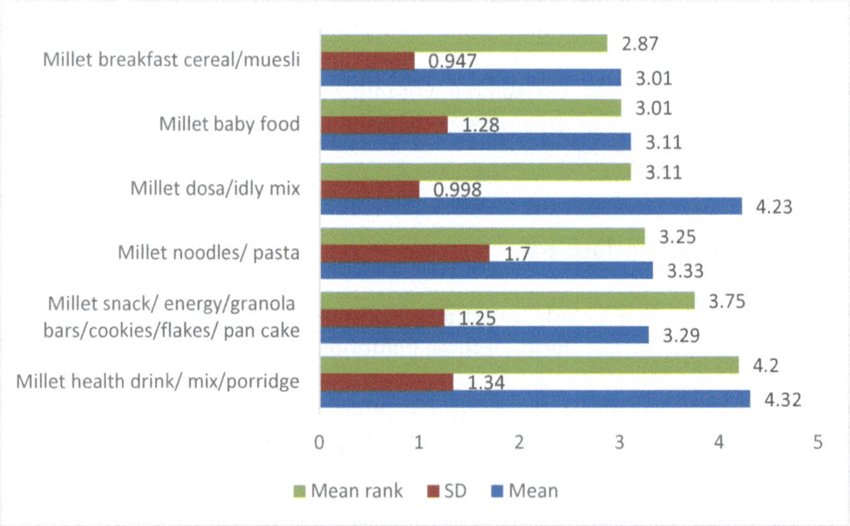

Fig. 16.2 Graphically representation of awareness and preference towards millet and millet-based products

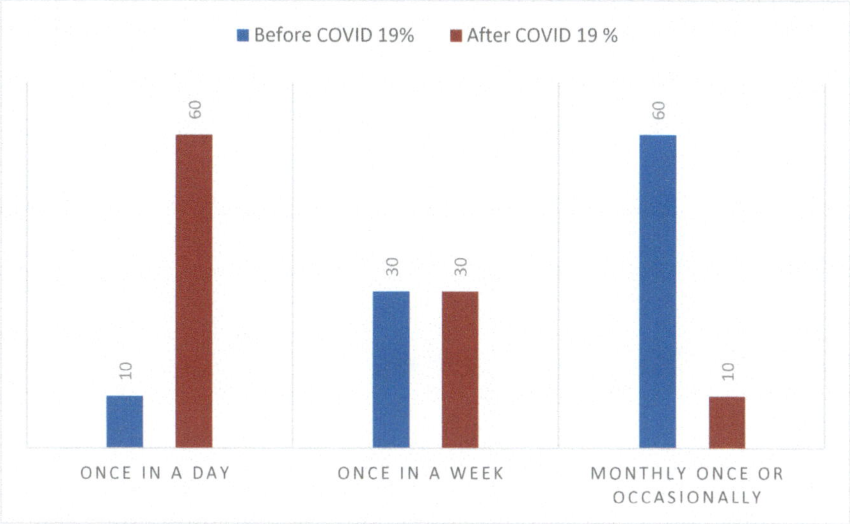

Fig. 16.3 Graphical representation of consumption of millets before and after COVID 19

- To appeal to the "variety seeking" mentality of customers, millet-based eateries should diversify their menus, particularly their online offers, in order to give a number of options for customers to choose from.
- Develop price methods that make millet-based goods accessible to a wide range of incomes, working in collaboration with government efforts or non-governmental organizations to achieve the goal of making them more affordable.

References

Anagha, K. K. (2023). Millets: Nutritional importance, health benefits, and bioavailability: A review. *Energy, 329*(328), 361.

Samtiya, M., Aluko, R. E., Dhaka, N., Dhewa, T., & Puniya, A. K. (2023). Nutritional and health-promoting attributes of millet: Current and future perspectives. *Nutrition Reviews, 81*(6), 684–704.

Tripathi, G., Jitendra Kumar, P. H., Borah, A., Nath, D., Das, H., Bansal, S., et al. (2023). A review on nutritional and health benefits of millets. *International Journal of Plant & Soil Science, 35*(19), 1736–1743.

Millets and Its Value-Added Products 17

Angshumitra Banerjee

Abstract

This paper delves into the nutritional value of millets and their multifaceted role in global nutrition and sustainability. Millets, a group of small-seeded grains, have garnered increasing attention due to their exceptional nutritional composition and potential to address contemporary challenges related to food security, health, and sustainable agriculture. Exploration begins with an introduction to the various types of millets and their significance in global nutrition. In the upcoming decades, the world agriculture production could decrease due to climate change. The role of millets as a smart food is inevitable. India is the leading producer of millets in the world with 10.9 million tons in the year 2019 (Das et al., 2019) . We can dissect the macronutrient and micronutrient profiles of millets, highlighting their rich content of carbohydrates, proteins, vitamins, minerals, dietary fiber, and antioxidants. In the culinary realm, we showcase the versatility of millets, both in traditional and modern recipes, transcending cultural boundaries and dietary preferences. We must emphasize the nutritional advantages of millet-based dishes and their incorporation into diverse cuisines worldwide.

Keywords

India · Processing · Nutrients · Dietary fiber · Importer · Modern life · Millet based product

A. Banerjee (✉)
The PSBB Millennium School Gerugambakkam, Chennai, Tamil Nadu, India

C. B. Singh et al. (eds.), *Global Millets Production for a Sustainable Future*,
https://doi.org/10.1007/978-3-031-88157-2_17

Introduction—I was concentrating on answer scripts correction, my daughter came with a bowl of snacks and asked me to have it. My mind was in correcting papers but I gave in and took a bite. My full mind reminisced the flavour and the known taste. I paid attention to see what I was eating and it was our Millet ladoo.

We started having a discussion about millets. Sanika asked me—"Ma do you know that we're celebrating International year of Millets in 2023?" I smiled and answer—"United Nation at the behest of Government of India in 2023 has declared International Year of Millets." She was impressed by my answer. I promised her to write an article on Millets.

The paper is organized in thirteen sections: (1) Introduction Millets and History, (2) Cultural importance of millets, (3) Millets Types and Name in Different Languages, (4) impact of millets (5) Millets Nutrition Facts per 100gms with rice and wheat (6) How do millets help farmers? (7) The largest importer of millets (8) The largest importer of millets (9) Can I eat millet every day? (10) Father of millet in Modern life (11) The millet lady of India (12) Where to buy and store (13) Conclusion.

17.1 Introduction Millets and the History

In India, millets have been mentioned in some of the oldest Yajurveda texts. It identifies foxtail millet (priyangava), Barnyard millets (aanava) and black finger millets (shyaamaka), thus indicating that millet consumption was very common, pre-dating to the Indian Bronze Age (4500 BC) (Fig. 17.1).

Fig. 17.1 Different types of millets

17.2 Cultural Importance of Millets

Millets have other important cultural and economic roles. In many parts of the world, millets are used in religious and cultural ceremonies and are also an important source of income for small-scale farmers.Millets are cereals from the Poaceae grass family. They are considered as one of the oldest cultivated crops. Generally, pearl millet (Pennisetum glaucum) and finger millet (Eleusine coracana) are known as the two major millets used for food and feed (Table 17.1).

17.3 The Impact of Millets

Millets are rich in non-starchy polysaccharides, fibre, and low glycemic index, which controls blood sugar levels, and are the ideal grain for diabetic patients. The soluble fibre and millet protein help to improve gut health and reduce cholesterol levels. It is a good source of protein, fiber, key vitamins, and minerals. The potential health benefits of millet include protecting cardiovascular health, preventing the onset of diabetes, helping people achieve and maintain a healthy weight, and managing inflammation in the gut etc. (Table 17.2).

17.4 How Do Millets Help Farmers?

Millet crops require low inputs; in addition to less water, millets require almost no chemical fertilizers or pesticides and can grow well in less fertile soil and harsh weather. Farmers do not have to worry about high investment and maintenance costs to produce millets.

17.5 The Largest Importer of Millets

Top Importing Countries-United States, Indonesia. United Arab Emirates, Vietnam, Netherlands,United Kingdom, Singapore, Germany.

17.6 Millet-Based Products You Must Try

Millet Flour, Millet Crackers, Millet Noodles:Millet Noodle Recipes for a Delicious Meal: Millet Noodle Salad: Millet Noodle Stir-Fry: Millet Noodle Soup,Millet Cereal, Millet Bread—cakes,Millet Drink,Millet Pasta, biscuits, muffins, crackers, laddoos and chikkis using different levels of incorporation in the form of flour or whole seeds (Fig. 17.2).

Table 17.1 Millets types and name in different languages

English	Pearl millet	Finger millet	Foxtail millet	Kodo millet	Little millet	Barnyard millet	Sorghum
Hindi	Bajra	Nachani, Mundua, Mandika, Marwah	Kangni,Kakum,Rala	Koden,Kodra	Kutki,Shavan	Jhangora,Sanwa	Jowar
Tamil	Kambu	Kezhvaragu,Kelvaragu ,Keppai,Ragi	Thinai	Varagu	Saamai	Kuthiravali (Kuthiraivolly)	Cholam
Telugu	Sajjalu	Ragula,Ragi Chodi	Korra	Arikelu,Arika	Sama,Samalu	Udalu, Kodisama	Jonna
Kannada	Sajje	Ragi	Navane	Harka	Saame,Save	Oodalu	Jola
Malayalam	Kambam	Panji Pullu	Thina	Koovaragu	Chama	Kavadapullu	Cholam
Marathi	Bajri	Nagli,Nachni	Kang,Rala	Kodra	Sava,Halvi,Vari	–	Jowari,Jondhala
Punjabi	Bajra	Mandhuka,Mandhal	Kangni	Kodra	Swank	Swank	Jowar
Gujarati	Bajri	Nagli,Bavto	Kang	Kodra	Gajro,Kuri	–	Jowari,Juar
Bengali	Bajra	Marwa	Kaon	Kodo	Sama	Shyama	Jowar
Oriya	Bajra	Mandia	Kanghu,Kangam,Kora	Kodua	Suan	Khira	Juara

Table 17.2 Millets nutrition facts per 100gms with rice and wheat

Crop/Nutrient	Protein (gm)	Fat (gm)	Fiber (gm)	Minerals (gm)	Iron (mg)	Calcium (mg)	Calories (kcal)
Pearl millet	10.6	4.8	1.3	2.3	16.9	38	378
Finger millet	7.3	1.5	3.6	2.7	3.9	344	336
Foxtail millet	12.3	4	8	3.3	2.8	31	473
Kodo millet	8.3	3.6	9	2.6	0.5	27	309
Little millet	7.7	5.2	7.6	1.5	9.3	17	207
Barnyard millet	11.2	3.9	10.1	4.4	15.2	11	342
Sorghum	10.4	3.1	2	1.6	5.4	25	329
Proso millet	12.5	2.9	2.2	1.9	0.8	14	356
Rice	6.8	2.7	0.2	0.6	0.7	10	362
Wheat	11.8	2	1.2	1.5	5.3	41	348

Fig. 17.2 Noodles and other products

17.7 Can I Eat Millet Every Day?

It is best to consume millets in moderate amounts; three or four times a week is ideal. Millets have numerous health benefit it helps to lower your blood sugar and cholesterol levels. They are also gluten-free, so people with celiac disease or gluten sensitivities can enjoy them.

However, it is not recommended for people with hypothyroidism. Few doctors has share, “People having thyroid disorders should avoid consuming millet because it leads to the enlargement of the thyroid.”

17.8 Father of Millet in Modern Life

Khadar Vali is known as the 'Millet Man of India'. He spoke about the nutritional benefits of millets at India Today Conclave in Mumbai. He said that there's no need for medicine if one incorporates millet into their diet. "When your food is right, there's no need for medicine."

17.9 The Millet Lady of India

Raimati Ghiuria is often called the "Queen of Millets" and rightly so. She has not only grown and preserved 30 varieties of rare millets but has also trained hundreds of women in cultivating these nutritious grains.

17.10 Where to Buy and Store

Millets are available in Organic shops and also in some super markets specially in India. Store millet in an airtight container in a cool, dry and dark place, where it will keep for several months.

17.11 Conclusion

Millets have an energy value similar to staple cereals. They provide more significant health benefits due to their high fiber, minerals, vitamins, macro- and micronutrients, and phytochemicals and can help combat chronic disorders. Making millets part of a regular diet can provide an affordable, complete, and healthy meal. It was observed that during germination and fermentation of millets, the dietary fiber, mineral, and vitamin content of most millets improved. Simple processing techniques such as soaking, germination/malting, and fermentation can help tackle the problem of protein–energy malnutrition by improving protein digestibility and the bioavailability of the minerals. However, it was observed that decortication, dehulling, milling, extrusion resulted in a reduction of total proteins, total dietary fiber, and micronutrients. Thus, we should care during the decortication of millets, as excessive dehulling can result in lower fiber content and loss of micronutrients due to the loss of nutrient-rich bran and germ portion. Awareness needs to be created at both commercial and household levels regarding the impact of processing methods on the nutritional properties of millets and the health benefits of millets.

Reference

Indian Khana, Botanical online, healthline, *The Pharma Journal*, National Library of Medicine, Google Browser.

Pearl Millet in India; Area, Production and Yield by Structural Stability Regression Model

18

V. Krishnan Kutty

Abstract

Pearl millet is a gluten-free grain that boasts a high fiber content and protein levels comparable to wheat. It is also abundant in antioxidants and vitamin B, contributing to improved digestion, heart health, and additional benefits. The quantity, area, and yield of pearl millet output, as well as its growth, trend, and structural stability, show an economy's stability and contribution. The purpose of this study is to compare pearl millet output, cultivation area, and yield per hectare in India before and after the new agricultural policy (1978–2000) in terms of growth, trend, and structural stability. (since 2001 to 2022). Secondary data from Agricultural Statistics at a Glance 2022 was used to fulfill the objectives. The trend line, t-test, and structural stability regression model were all used in the estimation process. From 1978–79 to 1999–2000, the overall growth rates of area, production, and yield of bajra were −41.17%, 72.7%, and 193.66%, respectively. The area increase is negative, but the output and yield of bajra are positive. The pattern reveals that the area of bajra agriculture is declining, yet production has changed slightly in pooling periods. There has been no structural change in the area under pearl millet cultivation and production as a result of India's new agricultural policy.

Keywords

Millet · Production · Regression · Agriculture

V. Krishnan Kutty (✉)
Government College Kodanchery, University of Calicut, Kozhikode, Kerala, India

C. B. Singh et al. (eds.), *Global Millets Production for a Sustainable Future*,
https://doi.org/10.1007/978-3-031-88157-2_18

18.1 Introduction

Pearl millet is the most extensively grown type of millet and is among the most drought-resistant domesticated cereals. Following its domestication, it rapidly spread throughout the semiarid tropical regions of Africa and Asia. India, known for its rich agricultural heritage, has been cultivating a diverse range of crops for centuries. Among these crops, millets have gained significant importance due to their nutritional value, adaptability to diverse climatic conditions, and contribution to sustainable agriculture. Millets are a highly varied group of small-seeded grasses widely grown around the world as cereal crops or grains for fodder and human food. In India, millet production has been on the rise in recent years. India is one of the largest producers of millets, and Indian farmers have been increasingly planting millet as a drought-resistant crop. The Indian government has also been promoting millet production as part of its National Food Security Mission. Millet is a type of grain that is popular in many parts of the world, especially in Africa and Asia. It is a staple food in many parts of the world, particularly in Africa and Asia. According to the World Food Programme, there are an estimated 1.2 billion people who consume millet as a part of their diet. Indian millets are a group of nutritiously rich, drought-tolerant crops mostly grown in the arid and semi-arid regions of India. They are small-seeded grasses belonging to the botanical family Poaceae. They constitute an important source of food and fodder for millions of resource-poor farmers and play a vital role in the ecological and economic security of India. These millets are also known as "coarse cereals" or "cereals of the poor." Indian millets are nutritionally superior to wheat and rice as they are rich in protein, vitamins, and minerals. They are also gluten-free and have a low glycemic index, making them ideal for people with celiac disease or diabetes. India is among the top 5 exporters of millets in the world. The world export of millet has increased from $400 million in 2020 to $470 million in 2021 (ITC trade map). India exported millets worth $64.28 million in 2021–22, against $59.75 million in 2020–21. The share of millet-based value-added products is negligible. India is the largest producer as well as the largest exporter of cereal products in the world. India's export of cereals stood at Rs. 96,011.42 crore, or 12,872.64 USD million, during the years 2021–22. Rice (including basmati and non-basmati) occupied the major share of India's total cereal exports, accounting for 75% (in value terms) during the same period. Whereas other cereals, including wheat, represented only a 25% share of total cereals exported from India during this period.

Millet production has remained relatively stable over the past few years, with an estimated production of 28 million metric tonnes in 2020. The majority of millet is produced in Africa, followed by Asia. India is the largest producer of millet, followed by Niger and China. Other major millet-producing countries include Burkina Faso, Mali, and Senegal. While millet is not a major food crop in the developed world, it plays a vital role in the diets of many people in developing countries. Millet is a drought-tolerant crop that can be grown in dry, arid climates where other crops would fail. It is also a nutritious grain that is high in fiber and essential minerals. For these reasons, millet will continue to be an important food crop in the years to come. India is one of the leading producers and suppliers of millet, and there are a number of millet sourcing points located throughout the country. The main millet-growing states in India are Rajasthan, Maharashtra, Karnataka, Andhra Pradesh, and Madhya Pradesh. These states have a large number of millet farmers who grow the grain for both domestic and international markets. In addition to the major millet-producing states, there are also a number of smaller millet-producing regions located throughout India. These regions include the states of Uttar Pradesh, Bihar, and Madhya Pradesh. The Indian government unveiled the New Agricultural Policy in July 2000. The government created this plan with the goal of encouraging the expansion and improvement of agricultural productivity and production, which will raise living standards, employment, and income levels. Supporting the general growth of the agriculture sector was the aim of this policy. The goal of the policy was to stimulate annual growth in the agriculture sector of more than 4%. Among the other goals are raising input productivity, raising value added per hectare, safeguarding the interests of small farmers, modernizing the agricultural sectors, halting environmental deterioration, promoting agricultural research and training, and getting rid of red tape. By supporting the commercially feasible, technically sound, ecologically benign, non-hazardous, and socially acceptable uses of the nations. Pearl millet, also known as Bajra, is among the most grown millets in India. It thrives in hot and dry conditions, making it suitable for cultivation in Rajasthan, Gujarat, Haryana, and Uttar Pradesh. Bajra is a splendid source of iron, calcium, and essential amino acids. Its coarse grains are used to make rotis, porridge, and even fermented beverages. The millets commonly grown in India include jowar (sorghum), bajra (pearl millet), ragi (finger millet), jhangora (barnyard millet), barri (proso or common millet), kangni (foxtail/Italian millet), kodra (kodo millet), etc. Of these, pearl millet, or bajra, was considered under study.

18.2 Objectives

To investigate the evolution and trends of Pearl Millet yield, area and production before and after the enactment of the new agricultural strategy.

To determine the level of dimensional stability in pearl millet production and the amount of land used for pearl millet cultivation.

18.3 Hypothesis

There has been no significant shift in agricultural production or area under pearl millet cultivation prior to or following India‘s new agricultural strategy.

18.4 Methodology

In order to achieve the study's goals of determining the production, cultivated area, and yield per hectare of pearl millet in India, we used secondary data. Data on the area under cultivation of bajra (in million hectares), production (in million tonnes), and oilseed yield per hectare (in kg per hectare) were provided by the Ministry of Agriculture and Farmers' Welfare of the Government of India in Agricultural Statistics at a Glance 2022. The data covered two time periods in India: Period I, which ran from 1978–1979 to 1999–2000, and Period II, which ran from 2001–2022, which corresponded to the time periods before and after the nation's new agriculture policy of 2000. Both time spans, for example, India's 1979–2022 timeframe, demand a careful examination. A structural stability regression model called the Econometric Model, Chou Test (Gregory Chou), is used to measure the structural stability of pearl millet in terms of both total production and area. At a 5% significance level, it is computed independently using a pooled sample from periods I and II. The production, area, and yield per hectare of pearl millets in India were estimated and compared using the average, compound annual growth rate (CAGR), ANOVA, and F test before and after the implementation of the new agricultural policy.

18.5 Literature Review

Pearl millet is a grain that is free of gluten and is rich in fiber, with protein levels similar to those found in wheat. It is also rich in antioxidants and vitamin B, which support better digestion, cardiovascular health, and other advantages. Pearl millet: A basic overview of an underutilized source of nutrition, Monika Satankar et al. (2020), Because it can grow in tough climate conditions and is high in nutrients, pearl millet is a significant grain crop. Unexpected climate shifts and other natural calamities can lead to issues with food security, driving up food prices and limiting the supply of food ingredients. Under these circumstances, pearl millet offers impoverished men an alternative nutrient-rich crop that supplies adequate nourishment for an active and healthy life. In comparison to other major cereal crops, it is an inexpensive source of nourishment. Despite the crop's nutritional value and health advantages, its use is limited because of anti-nutrition elements and poor storage quality. In 2021, Swati Kumari and others Production opportunity of pearl millet (Bajra): A case study in Bihar. The current study sought to determine the costs and returns, sustainability, and resource use efficiency faced by sample farmers in Bajra production. 120 sample farmers were chosen using a multistage sampling procedure from four villages in the Bihar district of Bhojpuri. Analysis tools included the modified multi-criteria approach, the Cobb-Douglas production function, cost concept A, and Garrett's ranking procedure. In response, sample farmers said that although bajra is more profitable than rice, it is not as sustainable. Pearl millet, bajra Shubhara Saraswat (2019), "The Millennium Food" One of the most significant grains and one of the oldest millets utilized by our ancestors is pearl millet (Pennisetum glaucum), also referred to as bajra. It grows in tropical and semi-arid regions of the world, mostly in Asia and Africa. It is renowned for having a high fiber content. In several Indian states, such as Rajasthan and Gujarat, it is a regular dinner. According to earlier research, pearl millets are a great way to get micronutrients like zinc and iron. Phytic acid and polyphenols are among the antinutrient factors and inhibitors found in pearl millet, which contribute to the limited bioavailability of iron and zinc.

18.6 Results and Discussion

Pearl millet thrives in regions with harsh conditions, including drought, poor soil fertility, low moisture levels, and elevated temperatures. It grows effectively in soils with high salinity or low pH levels. Due to its resilience in challenging growing environments, pearl millet can be cultivated in places where other cereal crops like maize or wheat would struggle to survive. This summer annual crop is particularly suitable for double cropping and rotational farming. The grain and fodder are important food and feed resources in countries such as Africa, Russia, India, and China. The structural stability of pearl millet's growth rates of output, area, and yield was investigated as a result of the new agricultural policy.

The area cultivated for bajra declined from 11.39 million hectares to 8.9 million hectares in period I with −21.86% growth rates, as Table 18.1 illustrates the area,

Table 18.1 *Area, Production and Yield of Pearl Millet - Period I*

Year	*Area* of bajra	Simple growth	*Production* of bajra	Simple growth	*Yield* of bajra	Simple growth
1978–79	11.39	–	5.57	–	489	–
1979–80	10.58	−7.11	3.95	−29.08	373	−23.72
1980–81	11.66	10.21	5.34	35.19	458	22.79
1981–82	11.78	1.03	5.54	3.75	470	2.62
1982–83	10.94	−7.13	5.13	−7.40	469	−0.21
1983–84	11.83	8.14	7.72	50.49	653	39.23
1984–85	10.62	−10.23	6.05	−21.63	569	−12.86
1985–86	10.65	0.28	3.66	−39.50	344	−39.54
1986–87	11.27	5.82	4.51	23.22	401	16.57
1987–88	8.71	−22.72	3.3	−26.83	378	−5.74
1988–89	12.04	38.23	7.78	135.76	646	70.90
1989–90	10.9	−9.47	6.65	−14.52	610	−5.57
1990–91	10.48	−3.85	6.89	3.61	658	7.87
1991–92	10.03	−4.29	4.67	−32.22	465	−29.33
1992–93	10.62	5.88	8.88	90.15	836	79.78
1993–94	9.55	−10.08	4.97	−44.03	521	−37.68
1994–95	10.22	7.02	7.16	44.06	700	34.36
1995–96	9.32	−8.81	5.38	−24.86	577	−17.57
1996–97	9.98	7.08	7.87	46.28	788	36.57
1997–98	9.89	−0.90	7.64	−2.92	773	−1.90
1998–99	9.3	−5.97	6.96	−8.90	748	−3.23
1999–00	8.9	−4.30	5.78	−16.95	650	−13.10

Source: Agricultural Statistics at a Glance 2022. Author Calculation

output, and yield of pearl millet. The yield per hectare of bajra increased from 489 kg/hectare to 650 kg/hectare with 32.92% growth, while the production of bajra cultivation changed from 5.57 million metric tonnes to 5.78 million metric tonnes with 3.77% growth. The greatest area and production simple growth rates were 38.23 and 135.76% in 1988–89, and the highest yield was 79.78% in 1992–93.

Table 18.1 illustrates the area, output, and yield of pearl millet in period II. With a growth rate of −31.84%, the area cultivated with bajra fell from 9.83 million hectares in period I to 6.7 million hectares in period II. The yield per hectare of bajra increased from 688 kg/hectare to 1436 kg/hectare with a growth rate of 108.72%. The production of bajra cultivation changed from 6.76 million tonnes to 9.62 million tonnes with a growth rate of 42.31%. In 2003–04, the area, output, and yield had the highest simple growth rates at 37.08, 156.57, and 87.05%, respectively.

Table 18.3 displays the average area, production, and yield of pearl millet for periods I and II. In pooled periods, the area had an average growth of −19.7%, while the production and yield of bajra had positive growth rates of 48.93 and 88.03%, respectively. Bajra millet yield, production, and area are all positively correlated; however, production is not positively correlated.

The compound annual growth rates of bajra cultivation in terms of area, yields, and production are shown in Table 18.4; these growth rates were negative for Periods I and II as well as for pooled periods. However, there were improvements in bajra yield and production.

Table 18.2 Production, Aarea and yield of pearl millet—Period II

Year	*Area* of bajra	Simple growth	*Production* of bajra	Simple growth	*Yield* of bajra	Simple growth
2000–01	9.83	–	6.76	–	688	–
2001–02	9.53	−3.05	8.28	22.49	869	26.31
2002–03	7.74	−18.78	4.72	−43.00	610	−29.80
2003–04	10.61	37.08	12.11	156.57	1141	87.05
2004–05	9.23	−13.01	7.93	−34.52	859	−24.72
2005–06	9.58	3.79	7.68	−3.15	802	−6.64
2006–07	9.51	−0.73	8.42	9.64	886	10.47
2007–08	9.57	0.63	9.97	18.41	1042	17.61
2008–09	8.75	−8.57	8.89	−10.83	1015	−2.59
2009–10	8.9	1.71	6.51	−26.77	731	−27.98
2010–11	9.61	7.98	10.37	59.29	1079	47.61
2011–12	8.78	−8.64	10.28	−0.87	1171	8.53
2012–13	7.3	−16.86	8.74	−14.98	1198	2.31
2013–14	7.81	6.99	9.25	5.84	1184	−1.17
2014–15	7.32	−6.27	9.18	−0.76	1255	6.00
2015–16	7.13	−2.60	8.07	−12.09	1132	−9.80
2016–17	7.46	4.63	9.73	20.57	1305	15.28
2017–18	7.48	0.27	9.21	−5.34	1231	−5.67
2018–19	7.11	−4.95	8.66	−5.97	1219	−0.97
2019–20	7.54	6.05	10.36	19.63	1374	12.72
2020–21	7.65	1.46	10.86	4.83	1420	3.35
2021–22	6.7	−12.42	9.62	−11.42	1436	1.13

Source: Agricultural Statistics at a Glance 2022. Author Calculation

Table 18.3 Average area, production and yield of pearl millet: Period I, period II, and pooled

	Pearl millet		
Measures	Area	Production	Yield
Average (Period I)	10.48	5.97	571.64
Average (Period II)	8.415	8.891	1074.864
Pooled Average	9.450	7.432	823.250
Variance (Period I)	0.922	2.291	21167.39
Variance (Period II)	1.296	2.598	57206.03
Pearson "r"	0.636	0.296	0.633

Source: Author Calculation

Table 18.4 CAGR of pearl millet area, production and yield—Period I, period II, and pooled

	Pearl millet		
Measures	Area	Production	Yield
CAGR (Period I)	−0.0117	0.0018	0.0136
CAGR (Period II)	−0.0181	0.0169	0.0357
Pooled CAGR	−0.0123	0.0128	0.0254

Source: Author Calculation

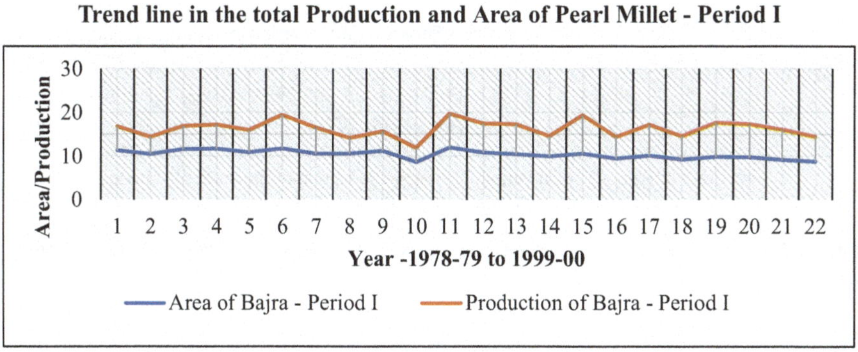

Fig. 18.1 Trend line in the total production and area of pearl millet—Period I. (Source: Author Calculation)

Figure 18.1 depicts the trend line for the total production and area of pearl millet during period I. The increase of area and production of bajra in India do not follow a similar pattern; instead, the area cultivated for bajra is trending downward, but the output is slightly different throughout period I.

Figure 18.2 depicts the trend line for both the total production and area of pearl millet during period II. In contrast, there is no comparable tendency for the expansion of bajra cultivation area and production in India; rather, the area under bajra cultivation is trending downward, albeit with some variation in production over period II.

The trend line in Fig. 18.3 depicts the area and total production of pearl millet during the pooled era. The area and production of bajra in India do not follow a

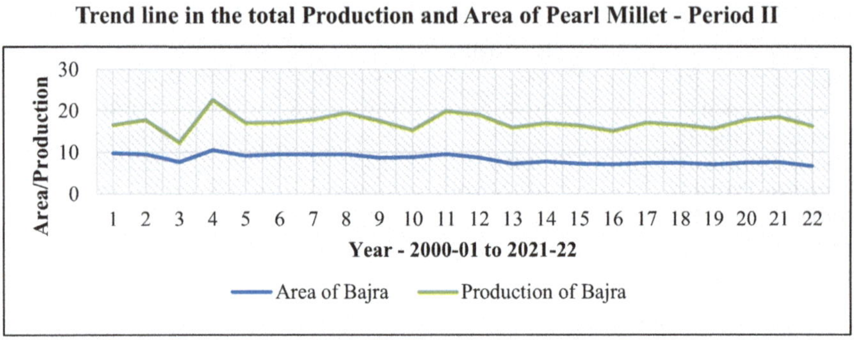

Fig. 18.2 Trend line in the total production and area of pearl millet—Period II. (Source: Author Calculation)

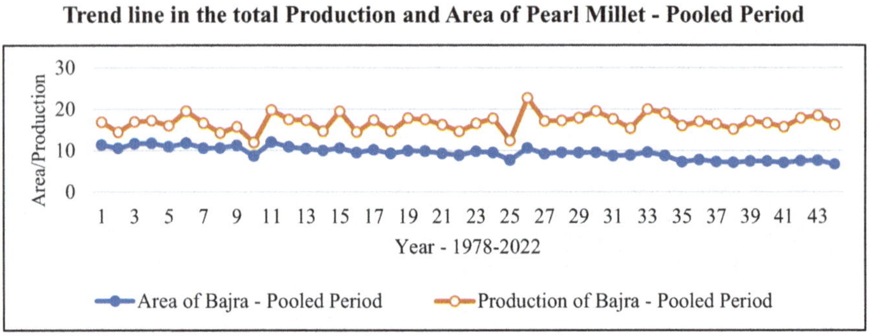

Fig. 18.3 Trend line in the total production and area of pearl millet—Pooled Period. (Source: Author Calculation)

Table 18.5 The structural stability of regression model—Production and area of pearl millet

Pearl millet	Pooled sample	Period I	Period II
	$W_t = 11.676–0.299M_t$	$W_{t(1)} = 9.75 + 0.12M_t$	$W_{t(2)} = 8.14 + 0.031M_t$
	$r^2 = 0.1881$	$r^2 = 0.037$	$r^2 = 0.0019$
	$N_1 = 76.049$	$N_2 = 18.632$	$N_3 = 27.171$
	D f = 42	D f = 21	D f = 21
	$N_4 = 45.803$ $N_5 = 30.246$ F = 13.207 P-value = 0.06		

Source: Author Calculation

similar pattern; instead, the area under bajra cultivation is trending downward, while the production varies little during the pooled periods.

Table 18.5 demonstrated the structural stability of the regression model in the area and production of pearl millet in India; no structural changes have occurred as a result of the country's new agricultural strategy in the area or production of pearl millet.

Table 18.6 ANOVA—Area, production and yield of bajra in period I & II

ANOVA—Area of bajra in period I & II						
Source of variation	*SS*	*Df*	*MS*	*F*	*P-value*	*F crit*
Between Groups	2,785,615	1	2,785,615	71.0857	1.410	4.072654
Within Groups	1,645,842	42	39186.71			
Total	4,431,456	43				
ANOVA—Production of bajra in period I & II						
Source of variation	*SS*	*Df*	*MS*	*F*	*P-value*	*F crit*
Between Groups	93.67364	1	93.67364	38.31389	2.12	4.072654
Within Groups	102.6858	42	2.4449			
Total	196.3595	43				
ANOVA—Yield of bajra in period I & II						
Source of variation	*SS*	*Df*	*MS*	*F*	*P-value*	*F crit*
Between Groups	2,785,615	1	2,785,615	71.0857	1.41	4.072654
Within Groups	1,645,842	42	39186.71			
Total	4,431,456	43				

Source: Author Calculation

The study of variations in pearl millet area under cultivation, production, and yield in periods I and II is presented in Table 18.6, which indicates that all P-values are more than 0.05, indicating that there is no significance in pearl millet area, production, or yield.

18.7 Conclusion

Pearl millet thrives in severe environments like drought, low soil fertility, low moisture levels, and high temperatures. It grows well in soils with high salt or low pH values. Pearl millet can be grown in areas where other cereal crops, such as maize or wheat, would struggle to survive. This summer annual crop is ideal for double-cropping and rotational farming. Africa, Russia, India, and China rely heavily on grain and fodder as food and feed resources. Millet's cultivation in India is extremely important because of its nutritional value, flexibility, and contribution to sustainable agriculture. The 10 types of millets discussed in this blog have a variety of health benefits. As people become more aware of the benefits of millets, their consumption is on the rise. Incorporating these adaptable grains into our diets can improve our general health and promote a sustainable food system. The overall growth rates of area, production, and yield of bajra were −41.17, 72.7, and 193.66%, respectively. The area increase is negative, but the output and yield of bajra are positive. The pattern reveals that the area of bajra agriculture is declining, yet production has changed slightly in pooling periods. There has been no structural change in the area under pearl millet cultivation and production as a result of India's new agricultural policy.

References

Kumari, S., Rahaman, S. M., Nayak, S., Vijay, K. V., Swami, S. (2021). Opportunity of pearl millet (Bajra) production: A case study in Bihar. *The Pharma Innovation Journal, SP-10*(10), 272–276.

Ministry of Agriculture and Farmers' Welfare of the Government of India in Agricultural Statistics at a Glance 2022.

Saraswat, S., & Krishna, S. (2019). Bajra (Pearl Millet) "The Millennium Food". *The Indian Journal of Nutrition and Dietetics, 56*(3), 325–334.

Satankar, M., et al. (2020). Pearl Millet: A fundamental review on underutilized source of nutrition. *Multilogic in Science, 34*(10), 1081–1084.

19 Analyzing Fluctuations in Millet Trends Across India: Unravelling Patterns in Area, Production, Productivity, and Consumption with a Focus on Growth and Instability

Shikha Singh and Sachin Rajas

Abstract

The pattern in the area, productivity, and result of finger millet, as well as the instability by Cuddy Della Valle list, have all been analyzed in the ongoing survey. India is one of the world's greatest makers of minor millets, and lately, the country's millets have become less rapidly. Little millets assume a significant part in areas of India that get weighty precipitation because of their capacity to endure droughts. It typically creates in unsatisfactory soil and environment conditions. Because of the wealth of excellent enhancements, minor millets are consumed in huge amounts in the north-eastern locale of India. As a result of social classes' consciousness of millets, consumption of millets is lower beyond the upper east. Eventually, kids' appetite in India is high. Individuals have relied upon an immense assortment of harvests starting from the beginning of human development, yet these harvests are today, as it were, turning out to be scant. A further outcome of our developing reliance on increasingly few yields is the weakening of our food supply. Millets are an exceptionally solid and delightful food that is consistently being dislodged by business crops. At the point when Cuddy and Della's instability record and coefficient of variation (CV) were utilized to analyze instability, they observed that both finger millet and pearl millet's drawn out productivity and production showed more extreme degrees of unpredictability. Compound annual growth rates (CAGR) exhibited that pearl millet productivity and production have expanded over the long haul notwithstanding a decrease in the area being worked on. Should the finger millet area being worked on experience a fall in yield while productivity increments.

S. Singh (✉) · S. Rajas
School of Management Ajeenkya D Y Patil University, Pune, Maharashtra, India

C. B. Singh et al. (eds.), *Global Millets Production for a Sustainable Future*,
https://doi.org/10.1007/978-3-031-88157-2_19

Keywords

Analyzing fluctuations · Millet trends · Unravelling patterns · Area · Production · Productivity · Consumption · Growth · Instability

19.1 Introduction

Millets are harder and stronger harvests that can flourish in unforgiving environments, areas with little precipitation, and lopsided soils. They likewise assume a crucial part in individuals' everyday weight control plans all through the vast majority of the world. Furthermore, millets have unrivalled dietary properties than different cereals like wheat and rice, and they support our country's food and wholesome security. As may be normal, it's as often as possible alluded to as coarse cereals; by and by, given the grains' overflow in supplements, it's presently called nutri-oats.

For quite a while, millet improvement has been the foundation of cultivating strategies in India, fundamentally reinforcing the country's food security and rustic business. Millets, which contain an unmistakable gathering of little developed grains like sorghum, finger millet, pearl millet, and foxtail millet, are striking for their nourishing advantages, flexibility in the climate, and expected job in practical planting. This exploration tries to lead a complete examination of the variations in millet trends all through India, digging into a few viewpoints, for example, the area a work in progress, productivity, production levels, and consumption trends. The key center is to unwind the intricate turn of events and instability patterns inside the millet area, offering knowledge into the inconspicuous elements that structure India's millet scene.

In India, millets have considered a cutting edge resurrection to be well as veritable congruity in their turn of events. Ordinarily tracked down in an assortment of agro-organic zones, millets have adjusted to a scope of climatic circumstances, showing flexibility to components like low-input cultivation and the dry season. Of late, developing familiarity with the nutritious worth of millets has incited their rebound, as purchasers, administrators, and ranchers perceive their capability to battle unfortunate eating and add to reasonable food frameworks. Nonetheless, it is basic to fathom the variations in millet trends to augment their true capacity and address difficulties achieved by moving green scenes and dietary propensities.

A fundamental piece of this examination is looking at the geological dispersion of millet improvement in India. Variations in agroclimatic conditions and cultivating rehearses across locales impact the determination of millet combinations and add to the variety saw in millet trends. This study means to distinguish center spots of millet advancement by isolating out these geographic patterns and analyzing the monetary variables and rustic practices that impact the spatial parts of millet production.

Likewise, the survey digs into the parts of millets' production, taking a gander at the elements impacting fluctuations in yield and result. Understanding the difficulties looked by millet farmers, for example, getting top notch seeds, giving reliable water system, and overseeing irritations, is urgent in concocting techniques that advance possible millet development. Besides, analyzing the occupation of government procedures, market components, and mechanical interventions in trim millet production trends would supply encounters into the more broadened setting of the millet area's extension and robustness.

One more significant part of this investigation is molded by the millet productivity appraisal. Through analyzing the yield per unit area, the review endeavours to recognize factors adding to changes in millet production, with an emphasis on down to earth rustic strategies, mechanical turns of events, and the reception of powerful developing procedures. Fathoming the elements that impact productivity can support the foundation of explicit go-betweens to improve millet advancement ability and guarantee that millet development is monetarily practical for smallholders.

This study means to give significant experiences that can explain arrangements, engage farmers, and advance the maintainable incorporation of millets into India's rural and wellbeing scene by analyzing the development and unpredictability inside the millet space.

19.1.1 Objectives of the Study

- To evaluate India's finger millet production, productivity, and area trends.
- To comprehend the elements influencing India's decline in minor millet cultivation.
- To assess the productivity and production instability of finger millet and pearl millet.

19.2 Literature Review

Agarwal and partners (2018) improve the text by leading an ethno-regular survey that spotlights on little millets in India. The audit underscores these grains' tremendous nourishing and contemporary potential. Through exhaustive examination of the different elements of little millets, the creators give important experiences into the assorted jobs these yields play in conventional and contemporary settings. The dietary importance demonstrated in this study ends up being especially significant in the worldwide setting, considering the developing accentuation on reasonable and sustaining food sources.

Balaji et al. (2017) add to the assortment of information by utilizing fundamental break examination to grasp little millets'production and consumption trends in India. This system strategy upholds the separating proof of significant variables impacting the components of millet improvement and consumption by considering a nuanced evaluation of trends. The discoveries hold importance for the two specialists and specialists, as they give a more profound comprehension of the developing scene of little millets in India's plant area.

Ahmed et al. (2022) move the concentration to the metropolitan climate, taking a gander at the qualities, causes, and results of dietary changes in Accra, which is quickly urbanizing. Through following the excursion from customary Ampesie to the advanced pattern of French fries, the audit catches the intricate transaction of financial elements affecting food decisions. The examination features the requirement for reasonable dietary decisions even with quick metropolitan change, adding to the bigger conversation on urbanization and its suggestions for food systems.

Chatenet et al. (2022), who cover everything from central standards to the freshest consistent strategies and contemporary turns of events. The survey incorporates accessible information and gives goodies of data about arising forward leaps and their expected effect on reasonable energy production. The mix of data assembled from this survey fills in as an important asset for specialists, chiefs, and finance managers anxious to advance water electrolysis as a fundamental part of the ideal energy progress.

Das and Gundimeda (2022) add to the writing by analyzing the adequacy of extending the utilization of biofuels in emerging nations. The audit reveals insight into the perplexing connections among land use, food security, and biofuel production through an assessment of exact proof. The discoveries feature the requirement for a nuanced way to deal with biofuel upgrade that considers its natural and prudent outcomes on area organizations, and they offer ideas for system specifying.

Fridah (2021) looks at how land-use patterns among the Kabras in Kenya's Kakamega Area have changed somewhere in the range of 1963 and 2010. The doctorate show presents an observationally upheld viewpoint on how land-use patterns have been formed over the course of time by outside impacts, country propensities, and monetary elements. The examination expands how we might interpret the perplexing elements that impact land-use changes and gives experiences into the sort of viable land that the board rehearses.

Ibraheem et al. (2018) use sorghum as a model collect and spotlight on utilizing omics forward leaps to unravel plant pressure responses. The paper sums up the headway made in figuring out the crucial tension responses of subatomic parts. This overview not just reveals insight into late improvements in the field of plant science, yet it likewise underscores the meaning of utilizing model yields, like sorghum, to upgrade how we might interpret pressure change in crops, which is pivotal for worldwide food security.

Karamchedu (2023) digs profoundly into the political economy of groundwater-driven agrarian improvement in dryland India. The paper explores the complicated communication of regulative worries, monetary issues, and natural perspectives in delivering agrarian settings. Through an assessment of the job that groundwater plays in green change, the creator reveals insight into the difficulties and open doors related with reasonable water the board and rural advancement in water-centered areas.

Kumar et al. (2023) A complete examination of ecological dangers and their association with food openness in India is given by at the confined level. The audit, which is the aftereffect of ICRISAT and WFP cooperating, evaluates what ecological dangers mean for food security. The discoveries shed vital light on the helplessness of food frameworks to ecological change and give proof based ways to deal with naturally adaptable horticulture and confined food availability.

19.3 Millet Cultivation Landscape in India

India's millet improvement scene incorporates a different and boundless cultivation setting, mirroring the country's rich agroclimatic conditions and cultural variety. For quite a while, millets—which incorporate sorghum, finger millet, pearl millet, foxtail millet, and others—have been a fundamental piece of Indian farming. Topographically, the growth of millet happens in a few states, with explicit geographic trends and changes because of soil organization, precipitation patterns, and temperature fluctuations.

Pearl millet, or bajra, is a staple yield in the completely dry and semi-dried districts of western and focal India because of its versatility to cruel circumstances. Essentially, finger millet, or ragi, is esteemed in the southern states in light of the fact that to its healthy benefit and flexibility to fluctuating territory. Then again, sorghum tracks down favor in tropical and semi-completely dry conditions.

A valid outline of millet's growth in India uncovers the harvest's significance in conventional cultivating frameworks. Millets were regularly developed as precipitation oversaw crops, assuming a significant part in decreasing the gamble related with capricious precipitation. These harvests have been pivotal to the occupations of little and unimportant farmers, regularly filling in as a solid wellspring of food and feed.

In any case, the millet improvement situation has been affected by shifts in rural strategies and food inclinations. Millets' growth was stifled by the Green Surprise, which advanced high-yielding assortments of wheat and rice. As of late, millets have been the subject of recharged research because of their expected dietary benefit, flexibility to different natural circumstances, and attainability for plant application.

These harvests have returned thanks to government drives and care crusades that are advancing millet improvement. The improvement scene is going through change as farmers utilize progressed advancement methods, utilize regular development practices, and examination market potential open doors for millet-based items.

Understanding India's millet improvement scene is fundamental for practical agribusiness and food security. It involves dealing with common abberations, advancing areas of strength for climate, and incorporating millets into traditional cultivation strategies to upgrade their true capacity for improving the general flexibility of the farming locale.

19.4 Deciphering the Dynamics of Area, Production, and Productivity

A broad assessment of the confusing perspectives impacting these pivotal country limits is important to decipher the components of area, production, and productivity concerning millet improvement in India.

- **Area Trends**

To comprehend the geological scattering and regional tendencies of farmers, dissecting the patterns in the millet improvement area is fundamental. Variations being developed areas are brought about by different elements, for example, soil types, agroclimatic conditions, and conventional cultivating strategies. Geological subtleties assume a huge part, as various states or regions might show clear inclinations for explicit millet assortments because of neighborhood agro-natural circumstances. Moreover, farmers'choices about how much area to save for millet improvement may be impacted by monetary contemplations like as landholding patterns, water openness, and government drives.

- **Production Patterns**

The general gather result and the yield per hectare are two parts of millets' production. Various variables, like the reception of high-yielding millet collections, mechanical interventions, and the acknowledgment of worked on agrarian practices, impact the key measurement known as yield per hectare. Examining the impacts of natural instability, illness and vermin among leaders, and admittance to top notch information sources are extra moves toward understanding production patterns. Besides, the variety of millet collections made in various areas adds to changes in generally production levels.

- **Productivity Analysis**

The effectiveness of the agrarian cycle is reflected in the productivity of millet growth, which incorporates both the data yield proportion and the general presentation of the cultivating framework. Imaginative arrangements, like the utilization of further developed seed assortments, exact development strategies, and computerization, can fundamentally increment millet yield. Crop turn, intercropping, and composed administration rehearses are among the farmer errands that affect productivity levels. Understanding these parts involves assessing the monetary setting of cultivating networks, their admittance to assets, and their acknowledgment of useful horticultural strategies that advance long haul yield.

Basically, assessing the advancement of millet concerning area, productivity, and production expects for a sweeping methodology that takes topographical, conservative, and creative components into account. The establishment for organizing specific strategies, mediations, and techniques is laid out by this concentrate to progress valuable millet developing techniques and lift the flexibility of the rural district.

19.5 Methodology

Status refers to the current state of frightening circumstances or the status of the situation. The acreage, productivity, output, and consumption of individual millets were used to calculate the long-term status of millet. Finger millet and pearl millet were chosen for the current focus since they are the most commonly produced and farmed millets in India. Thus, optional data on the area, productivity, and production of finger millet and pearl millet were gathered between 2000–01 and 2016–17. Furthermore, additional information about millet production and home consumption in India was gathered between 1959–1960 and 2017–18.

19.5.1 Trend Analysis

The exceptional growth capacity was applied to evaluate the selected financial elements' growth rates, and the model is

$$Y = ab^t e \tag{19.1}$$

The formula y = (Against ln of b-1) X 100 was used to determine the percent compound growth rate (y).

Using the t logarithmic type in the case Y = stomach muscle e, the compound growth rate was determined as follows: Y = ln a + t ln b.

19.5.2 Instability Analysis

The coefficient of variation was used to concentrate on the differences in the area, production, productivity, and consumption of millets. The coefficient of variation (CV) was recorded using the accompanying recipe.

$$CV = \frac{Standarddeviation}{Mean} \times \mathbf{100} \tag{19.2}$$

The initial time series data on area, productivity, and production for the years 2000–01–2016–17, as well as the absolute production and consumption for the years 1959–60–2017–18, were fitted with a direct pattern.

Cuddy and Della's instability record was used to focus on the area's changeability as well as the output and productivity of millets:

$$CDVI = CV \times \sqrt{\mathbf{1} - R^{-2}} \tag{19.3}$$

CV = Coefficient of variation.

R^{-2} Differences in the multiple assurance coefficient The coefficient of variation was duplicated by the square base of the difference between the solidarity and the coefficient of different assessments (R^2).

19.6 Results and Discussion

The change in the entire area under pearl millet development between 2000–01 and 2016–17 was shown in Fig. 19.1. Over an extended period, the pattern showed that the area under pearl millet development had decreased from 9.83 million hectares to 7.47 million hectares. However, there was an upward tendency in India's pearl millet output from 2000–01 to 2016–17. From 6.76 million tons in 2000–01 to 9.8 million tons in 2013–06–17, pearl millet was produced worldwide. The reason for the increase in production is the sharp rise in pearl millet productivity, which increased from 6.9 quintals/hectare to 13.1 quintals/hectare despite a drop in the area under development between 2000–01 and 2016–17.

Figure 19.2 shows the gradual decrease in the area under finger millet development, from 1.76 million hectares in 2000–01 to 1.37 million hectares in 2016–17. Concurrently, productivity has declined from 15.5 quintal/hectare in 2000–01 to 13.5 quintal/hectare in 2016–17. Consequently, the overall production of finger millet fell throughout the same period, from 2.73 million tons in 2000–01 to 1.39 million tons in 2016–17.

Fig. 19.1 Trends in India's pear millet area, productivity, and production from 2000–01 to 2016–17

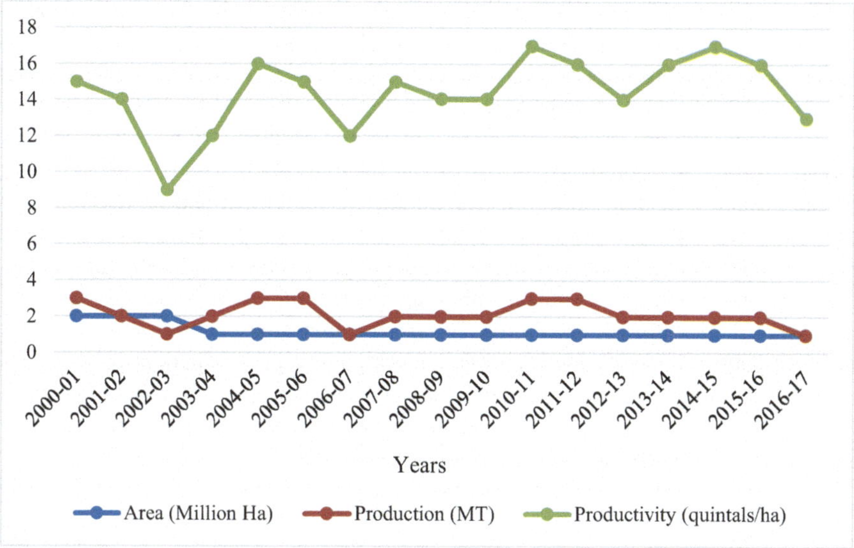

Fig. 19.2 Trends in finger millet area, productivity, and production in India from 2000–01 to 2016–17

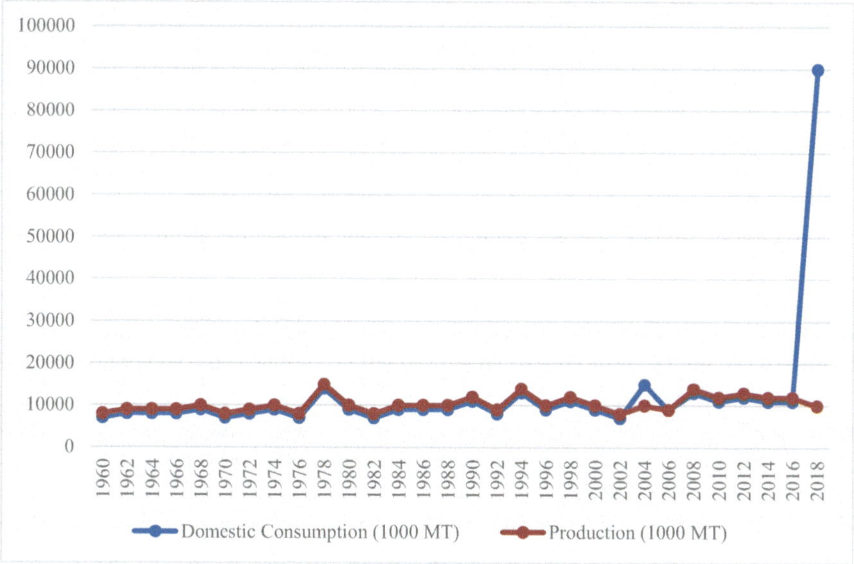

Fig. 19.3 Trends in India's total millet production and domestic consumption from 1960 to 2018

Figure 19.3 depicted the overall trend of millets' domestic consumption and total production in India from 1960 to 2017. The production and consumption lines' covering demonstrates how much millets are consumed domestically.

The overall trend in millets' production and consumption has been increasing.

The area's CAGR, finger millet and pearl millet productivity, and millets' overall production and consumption were broken down using their remarkable growth capacities. Table 19.1 demonstrates unequivocally that although pearl millet's development area has shrunk over time, output and productivity have nevertheless grown. Comparably, productivity has grown from 2000–01 to 2016–17 for finger millet, despite a decline in both the area under development and production over time. Between 1960 and 2017, the overall amount of millet produced and consumed domestically grew. The area's compound annual growth rates (CAGR) and the productivity and production of finger millet and pearl millet were broken down using the extraordinary growth capability. The pearl millet area of development shrank between 2000–01 and 2016–17, although productivity and production rose, according to an examination of the compound annual growth rate. Finger millet has caused a loss in development and production in the two regions under consideration, despite an increase in productivity between 2000–01 and 2016–17 (Fig. 19.4).

Table 19.2 appears to indicate that during the period (2000–01 to 2016–17), pearl millet output and productivity exhibited a more notable degree of volatility. However, compared to area and productivity throughout the same period, finger millet output shown a more significant degree of volatility (2000–01 to 2016–17). Between 1960 and 2017, there was a significant amount of fluctuation in both the total amount produced and the amount consumed domestically in millets. When it came to pearl millet, there was less variance observed in the area, and when finger millet was there, there was less instability observed in the case of a productivity event. A significant degree of heterogeneity was seen between homegrown consumption and absolute millet production. The productivity of pearl millet and finger millet, as well as their combined production, varied at a more notable rate between 2000–17. There was a significant amount of fluctuation in millets' total production and domestic consumption from 1959–60 to 2017–18 (Fig. 19.5).

The current study's findings demonstrated that although pearl millet's development area has shrunk over time, output and productivity have nevertheless increased. In a similar vein, domestic millet production and consumption have grown over time despite increases in productivity and area under development for finger millet. Upon examining the instability of area, production, productivity, and consumption,

Table 19.1 Area, productivity, and production growth of finger millet and pearl millet from 2000–01 to 2016–17, as well as the overall output and domestic consumption of millets from 1960–2017 (in percentage terms)

Crop	Parameter	CAGR
Pearl millet	Area	−1.80
	Production	1.67
	Productivity	3.39
Finger millet	Area	−2.27
	Production	−1.55
	Productivity	1.34
Millets	Total production	0.615
	Domestic consumption	0.639

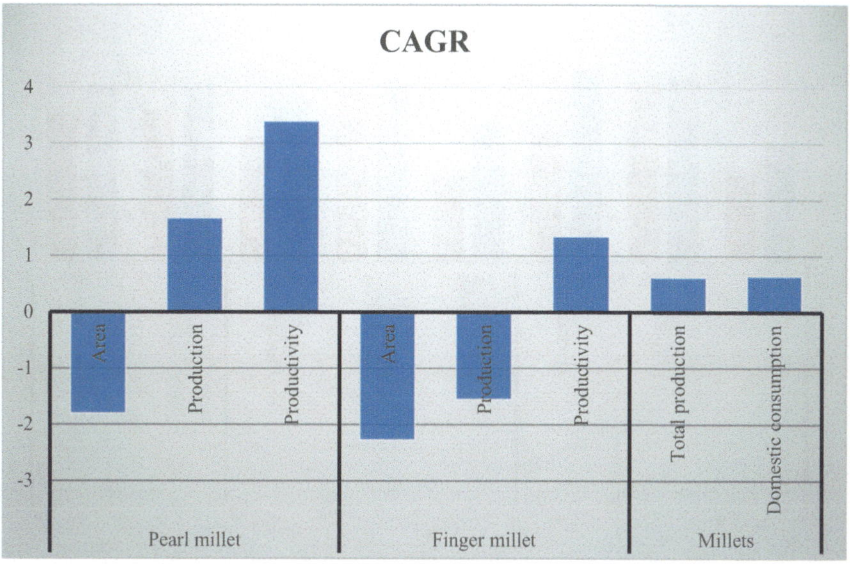

Fig. 19.4 The graph shows the percentage changes in the overall production and domestic consumption of millets from 1960 to 2017 as well as the growth in area, productivity, and production of finger millet and pearl millet from 2000–01 to 2016–17

Table 19.2 Variations in total millet production and domestic consumption from 1960 to 2017 (In %) as well as in the area, productivity, and production of finger millet and pearl millet from 2000–01 to 2016–17

Crop	Parameter	CV	CDVI
Pearl millet	Area	12.94	9.53
	Production	20.24	19.70
	Productivity	21.24	14.36
Finger millet	Area	15.28	9.78
	Production	20.16	18.79
	Productivity	14.63	13.72
Millets	Total production	19.58	16.68
	Domestic consumption	19.39	16.32

it was found that the production of finger millet and pearl millet had a greater degree of volatility in both output and productivity when compared to area and productivity. There was a noticeable difference between the incidence of homegrown eating and commercial millet production.

There are numerous possible explanations for the decrease in the area developed with millets. A focus on creating input-responsive, high-yielding assortments of grains, such as wheat and rice, was placed during the green transformation. When given improved technology, data sources, and motorization, wheat and rice demonstrated a significant increase in productivity and output. That being said, compared to more advanced varieties of wheat and rice, native millets are less information-responsive and yield less. Millets were thus replaced by those more lucrative

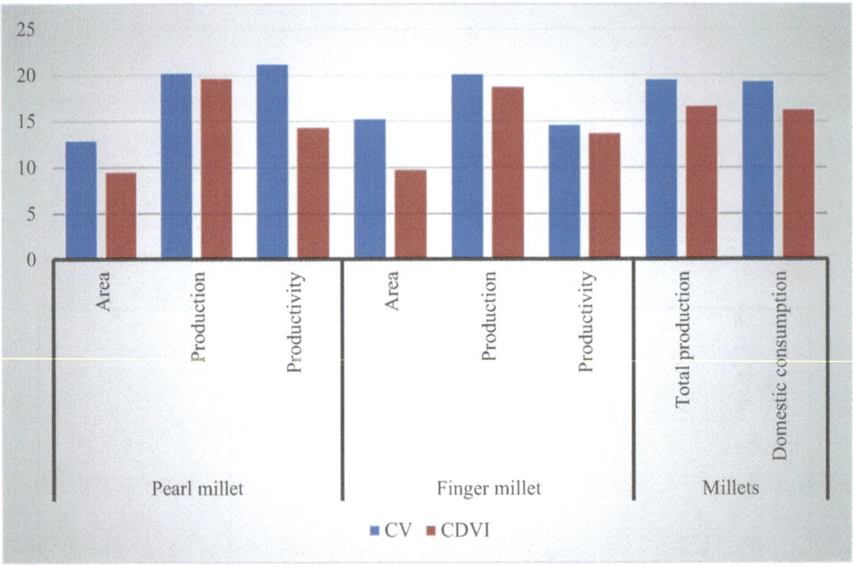

Fig. 19.5 A graph showing the variations in the area, productivity, and production of finger millet and pearl millet from 2000–01 to 2016–17, as well as the total millet production and domestic consumption from 1960–2017 (in %)

harvests when ranchers were granted access to manures, agro-synthetic substances, and high-yielding varieties of wheat, rice, and other business crops.

Additionally, the majority of our country's asset-poor little and minimal ranchers are the ones that cultivate millets and are also the primary consumers of millets. However, over time, millets have been referred to be "unfortunate man's food" and "substandard food" in comparison to wheat and rice, with the belief that these foods are consumed by the poor. Due to this inequality, people who make more money have shifted to eating wheat and rice, which has decreased the demand for millets and the amount of land under development, both of which have steadily decreased since Lentz, C.'s discovery.

19.7 Conclusion and Recommendation

An intricate understanding of the intricate patterns shaping the elements of the millet area has been provided by the analysis of variations in millet trends throughout India. Through examining critical limits such as the area under development, productivity, output levels, and consumption patterns, this study has revealed significant insights into the topographical, farming, and financial factors that impact the development of millet. In addition to providing a broad overview of millet production and domestic consumption, the evaluation of growth and instability in finger millet and pearl millet has highlighted the two challenges and possible opportunities for the region. The notable trends provide a framework for designated intercessions, emphasizing the adoption of innovative ideas, market integration, and reasonable rural practices to increase the adaptability and viability of millet farming. According to the review, while the area under development for pearl millet has decreased over time, productivity and output have increased nonetheless. Similarly, while productivity has increased for finger millet, both the area under development and production have decreased over time. Over time, both the total amount of millet produced and the amount consumed domestically have increased. An analysis of instability revealed a more significant degree of variance in the yield and production of finger millet and pearl millet.

In order to revitalize India's millet region, specific farming plans should tackle issues such as seed availability and irrigation. Spreading innovation, connecting markets, and promoting consumer awareness are essential for economic success. Support from the government for creative efforts to produce robust millet assortments is essential for adapting to changing conditions. In order to improve food security, advance sustainable farming, and address provincial occupations, coordination of these measures is essential.

References

Agarwal, S., Singh, A., Vivek, C., & Vikash, C. (2018). Ethno botanical study of small millets from India: Prodigious grain for nutritional and industrial aspects. *International Journal of Chemical Studies, 6*(4), 2155–2162.

Ahmed, A., Lazo, D. P. L., Alatinga, K. A., & Gasparatos, A. (2022). From Ampesie to French fries: systematising the characteristics, drivers and impacts of diet change in rapidly urbanising Accra. *Sustainability Science*, 1–25.

Balaji, S. J., Anbukkani, P., & Nithyashree, M. L. (2017). Production and consumption of minor millets in India – A structural break analysis. *International Journal of Agricultural Sciences, 38*(4), 1–8.

Chatenet, M., Pollet, B. G., Dekel, D. R., Dionigi, F., Deseure, J., Millet, P., et al. (2022). Water electrolysis: from textbook knowledge to the latest scientific strategies and industrial developments. *Chemical Society Reviews, 51*(11), 4583–4762.

Das, P., & Gundimeda, H. (2022). Is biofuel expansion in developing countries reasonable? A review of empirical evidence of food and land use impacts. *Journal of Cleaner Production, 372*, 133501.

Fridah, A. I. (2021). *The changing patterns of land-use among the Kabras of Kakamega County, Kenya C. 1963-2010*. Doctoral dissertation, Doctoral Dissertation, Kenyatta University.

Ibraheem, O., Adigun, R. O., & Olatunji, I. T. (2018). Omics technologies in unravelling plant stress responses; using sorghum as a model crop, how far have we gone. *Vegetos, 31*, 2. https://doi.org/10.4172/2229-4473.1000405

Karamchedu, A. (2023). Creating the 'Rice Bowl of India': Examining the political economy of groundwater-led agrarian transformation in dryland India. *Environment and Planning E: Nature and Space, 7*(3), 25148486231217141.

Kumar, S., Khan, P. E., Das, A., Pramanik, S., Nedumaran, S., Padmanabhan, J., et al. (2023). *ICRISAT and WFP: India Working Paper Quantifying Climate Hazards and Its Relationship with Food Availability: A Comprehensive District Level Analysis in India* (Working Paper). World Food Programme.

Index

C. B. Singh et al. (eds.), *Global Millets Production for a Sustainable Future*, https://doi.org/10.1007/978-3-031-88157-2

The manufacturer's authorised representative in the EU is Springer Nature Customer Service Centre GmbH, Europaplatz 3, 69115 Heidelberg, Germany. If you have any concerns regarding our products, please contact ProductSafety@springernature.com

Printed and bound by CPI Group (UK) Ltd, Croydon, CR0 4YY
30/09/2026
02225572-0009